Number Sense
and Number Nonsense

Number Sense and Number Nonsense

Understanding the Challenges of Learning Math

by

Nancy Krasa, Ph.D.

and

Sara Shunkwiler, M.Ed.

·P·A·U·L·H·
BROOKES
PUBLISHING CO ®

Baltimore • London • Sydney

Paul H. Brookes Publishing Co.
Post Office Box 10624
Baltimore, Maryland 21285-0624
USA

www.brookespublishing.com

"Paul H. Brookes Publishing Co." is a registered trademark
of Paul H. Brookes Publishing Co., Inc.

Typeset by Integrated Publishing Solutions, Grand Rapids, Michigan.
Manufactured in the United States of America by
Sheridan Books, Inc., Chelsea, Michigan.

The illustrative cases in this book are used by permission;
names have been changed to protect confidentiality.

Library of Congress Cataloging-in-Publication Data

Krasa, Nancy.
    Number sense and number nonsense : understanding the challenges
of learning math / by Nancy Krasa and Sara Shunkwiler.
        p.   cm.
    Includes bibliographical references and index.
    ISBN-13: 978-1-59857-020-5 (pbk.)
    ISBN-10: 1-59857-020-X (pbk.)
    1. Number concept.   2. Number concept in children.   I. Shunkwiler, Sara.
II. Title.

QA141.15.K73 2009
372.7—dc22   2009004895

British Library Cataloguing in Publication data are available from the British Library.

2012   2011   2010   2009
10   9   8   7   6   5   4   3   2   1

Contents

About the Authors

Nancy Krasa, Ph.D., is a practicing clinical psychologist with more than 25 years' experience in psychological, neuropsychological, and psychoeducational evaluation. She has served on the adjunct faculties of the colleges of medicine at Cornell University, New York University, and The Ohio State University. Dr. Krasa is a member of the International Dyslexia Association and has published articles in the fields of psychiatric and psychoeducational diagnosis. She received her bachelor of arts degree in mathematics from Smith College and her doctorate in clinical psychology from New York University. She and her husband live in Columbus, Ohio, and have three grown children.

Sara Shunkwiler, M.Ed., taught middle school at Marburn Academy, a private school in Columbus, Ohio, for bright children who learn differently. She is currently teaching Pre-Algebra and Algebra in a public school setting and has worked with many students who have varying degrees of math difficulty. Prior to entering teaching, she was an engineer, a career she chose when she placed third in a schoolwide algebra contest and discovered she was "good at math." Her team won the state competition, and the three women on that team went on to graduate at the top of their engineering classes at The Ohio State University. Ms. Shunkwiler also earned a master of science degree in ceramic engineering from the University of Illinois at Urbana-Champaign. For 12 years, she worked as a product development and test engineer with General Motors and received three United States patents during that time. She left engineering to share her love of mathematics and science with students in the pivotal middle school years, and she earned a master of education degree in middle childhood mathematics and science education from The Ohio State University. She and her husband, also a ceramic engineer, live in Frederick, Maryland, with their two teenage sons.

Preface

In the early fall a few years ago, a college senior named Abby showed up in tears to my psychology practice for a diagnostic evaluation. She was a hard-working honor student and respected peer tutor in English, but there was a good chance that she would not graduate. Why? To receive a degree from her college, she was required to pass one class of pre-calculus level mathematics.

Abby had struggled with math throughout her elementary years and, even though she did well in her other subjects, was barely able to earn enough math credits to graduate from high school. An exam during her college orientation had placed her into a non-credit remedial math course. After four failed attempts to pass that class, the dean referred her for an evaluation. She was frantic by the time she arrived at my office.

What was I to make of Abby? Her predicament raised many questions. Was her math difficulty a symptom of a disorder, as suggested by official psychiatric guidelines? If so, what evidence was I looking for? As a psychologist, I had seen many students who did poorly in math—in fact, many students seeking evaluations, for whatever reason, complain of struggling with math. That these students had trouble with math was indisputable; my job was to figure out *why*—without knowing the source of the problem, there would be no way to fix it. Psychiatric guidelines suggest that various cognitive impairments can be involved. Indeed, each student, including Abby, demonstrated a unique set of cognitive strengths and weaknesses. There was no pattern in their cognitive test results that might explain what they all had in common, which was math failure. Nor was there any clear way to understand how their individual cognitive weaknesses contributed to this shared outcome.

I also wondered whether Abby's plight was qualitatively different from other students' math challenges. Perhaps it was just the extreme end of a continuum that embraced any number of humanities majors who were never referred for evaluation simply because they were not required to take math. Many people, like Abby, do not like numbers and, when possible, avoid balancing their checkbooks, calculating discounts, and doubling recipes. And what about younger students who struggle with basic math until they are allowed to drop it, or who still need a calculator to reckon 4×6, even after years of flashcards? How prevalent are such difficulties? Some evidence suggests math troubles are pervasive: A recent Google search of the phrase *I suck at math* turned up 53,000 hits, mostly message-board math queries. How could I help Abby and others like her make sense of their math difficulties, and what advice could I give to their teachers and professors?

And so my quest began. One math curriculum expert told me that there was little research on math learning disabilities, and in a strict sense she was right. A few large-scale studies have shown, for example, that some students with math difficulties also have a reading disorder, but that many do not, and the few existing neuropsychological studies have been inconclusive; that was about all the available information. Moreover, one prominent researcher told me that science was not close to producing assessment or educational guidelines. Indeed, in that regard, research on math impairments is at least a decade behind that on dyslexia.

It quickly became clear that to understand why some people fail in math, I first had to understand what enables most people to succeed. That is, what cognitive skills are necessary for learning basic mathematics? On this topic, as it turns out, there is abundant research. Indeed, reviewing it has been, to borrow a phrase from Geekspeak, a bit like drinking from a fire hose. Scientists in animal and infant behavior, cognitive psychology and development, education, linguistics, genetics, neuropsychology, and most recently, neuroscience have all made preliminary though significant contributions to the field. Unfortunately, the research has largely remained buried in the journals of the individual academic disciplines; consequently, these findings have been inaccessible to educators, psychologists, and even to researchers across fields. Those results that have seen daylight have emerged in books for the general reader who is curious about the mind and mathematics, in texts restricted to arithmetic development in very young children, and more recently, in a few scholarly essay collections. For this reason, I decided to write a book that would integrate, for the first time, this vast body of work for practitioners who have a stake in its contributions.

How one learns math, however, depends on what happens in the classroom, as well as on what happens in the mind. Solving that piece of the puzzle would require the perspective and insights of a teacher both well versed in math and mathematical pedagogy and experienced in teaching students with learning differences. Hence, I recruited Sara Shunkwiler to join the project. A middle school math teacher with a special interest in children with learning disabilities, she knew first-hand the frustration of teaching students who, despite their hard work and her own dedication, simply could not master even the most basic material. She readily accepted my invitation; in particular, she was eager to make the psychological insights found in the research accessible to the classroom teacher and to address several broad, pertinent educational issues. To this end, she wrote the book's final chapter and additionally provided invaluable advice on the other ten.

Together, then, we aim to fill the literature gap with a book that asks the following questions: What cognitive skills are necessary for doing mathematics and how do those skills develop? Why do some people have trouble with math and how can that difficulty be evaluated? What do the scientific findings to date imply for education? Where should research go from here? We intend to answer these questions based on the latest scientific data, which in most areas are still quite preliminary.

Here are our arguments. We find that mathematics draws on three basic modes of thought and extraordinarily complex brain circuits, comprising a wide variety of related perceptual, cognitive, executive, and reasoning skills that under ideal conditions work together seamlessly. Because these systems depend on each other for full and efficient functioning, an impairment or perturbation in any part of the network can interfere significantly with the ability to learn and do mathematics. Math disability, therefore, is not unitary. Rather, it appears to result from any number of cognitive glitches, impairments, and asymmetries, often exacerbated by emotional and cultural issues or by instruction poorly matched to the student's way of thinking. For this reason, any evaluation of a student for unexpected and debilitating difficulty in learning math must be comprehensive. Math disability also exists on a continuum with ability such that the severe difficulties of some individuals are not qualitatively different from the occasional troubles experienced by the rest of us. Thus, the scientific findings pertain not just to students with severe and pervasive math disabilities but to anyone who struggles with or feels uneasy around numbers.

In addition, we argue that mathematics is difficult because the relevant cerebral networks, with one partial exception, are not inherently specific to number. Mathematics must be learned, and teachers will be most effective when they understand their students' unique intellectual strengths and weaknesses. A number of pedagogical tools and techniques already on the market and in the classroom appear compatible with the scientific evidence, but their effectiveness—for either the general classroom or for struggling students—has only recently begun to be subjected to rigorous testing. Other simple techniques have proven surprisingly effective. We review herein the handful of recent pedagogical studies.

Our goal is to make the science of mathematical cognition accessible to a variety of professionals, including both practitioners and academics. For practitioners—teachers, psychologists, and other professionals in daily contact with confused and discouraged students—we have synthesized the scientific findings in plain English, using the technical terms of the academic disciplines when necessary, but supplying definitions as well as examples and illustrations for additional clarification. By including illustrative student cases, we hope to bring the research findings to life and to convey the toll that our lack of knowledge takes on frustrated students and, by extension, on the lives of adults and on the economy and productivity of the nation.

For academics—those doing the research—we provide a wide-ranging bibliography, our gift to graduate students everywhere. Our review focuses chiefly on controlled studies, citing other work only when its insights seem especially pertinent. We alert all readers at the outset, however, that many of the studies are small and narrowly focused; most need to be replicated. Although many are excellent, some are methodologically flawed and thus should be interpreted cautiously. This book was never intended to be the last word on the subject—indeed, it is one of the very first words—and was written partly to provide researchers with a starting point for their own future work.

In attempting to write a book that would be accessible to a diverse readership, we elected to locate the scholarly apparatus in chapter endnotes. The advantage, as we see it, is to make the text more readable by removing disruptive in-text citations. Moreover, doing so left us free to cite more liberally, a benefit to academics wishing to mine the sources.

As recently as 2006, American 15-year-olds scored below average in mathematics among the world's most prosperous countries.[1] Thus this book should interest not only teachers, psychologists, other practitioners, aspiring professionals, and researchers, but also public policy officials and those whose curiosity is sparked by the national debate over how to improve American mathematics education.

Nancy Krasa

Note

[1]Organisation for Economic Co-Operation and Development, 2007, Table 6.2c.

Acknowledgments

Writing a book such as this one is humbling. Although our names appear on the cover, the real authors are to be found in the Bibliography at the back: the hundreds of scholars whose labors are at the heart of this story. Our job was merely to sort, weigh, and synthesize their findings (and any errors in that effort are entirely our own). So first, we would like to extend our appreciation to them and express the hope that they continue their good work.

Questions occurred to us as we combed through the material, and one of the great pleasures in writing the book was having the opportunity to correspond with those on the research front lines. We are enormously grateful to the scholars and writers who took the time to respond to our queries: Wendy Ark, Scott Baker, Miriam Bassok, Michele Berg, Olessia Blajenkova, Ellen Boiselle, Randee Booksh, Julie Booth, Linda Brody, Brian Butterworth, Ben Clarke, Stanislas Dehaene, Chris Donlan, Greg Duncan, Skip Fennell, Dawn Flanagan, Ann Gallagher, David Geary, Wim Gevers, Adam Gopnik, Sharon Griffin, Sherby Jean-Leger, Phil Johnson-Laird, Linda Jones, Vladik Kreinovich, Linda Liu, David Lubinski, Nicole McNeil, Kevin Miller, Istvan Morocz, Nora Newcombe, Michael O'Boyle, Bonamy Oliver, Thad Polk, Chris Robichaux, Joshua Rosenbloom, Amy Shelton, Margaret and Fred Stern, David Stevens, Claudia Uller, David Uttal, Michael von Aster, John Webb, and Jingzhu Zhang.

In addition, we extend a special and heartfelt thanks to a few scholars who gave so generously of their time to meet with us, read drafts, or answer lengthy questions: Yulia Kovas, Kelly Mix, John Opfer, Ruth Shalev, and Robert Siegler. Their advice and encouragement meant a great deal to us.

We would like to thank Marburn Academy—an independent day school in Columbus, Ohio, for children with learning differences—for their interest in our project. In particular, we extend our appreciation to Headmaster Earl Oremus, counselor Kelly Hunter-Rice, math director Valerie Blom, and the devoted teachers and staff for their time and enthusiasm. We especially thank the families, from Marburn and elsewhere, who agreed to let us tell their stories.

In ways large and small, family, friends, and colleagues contributed their time and invaluable advice. This book could not have been written without them, and we are deeply grateful: Deborah Aiges, Kathy Aitken, Shelley Austrian, Anna Balas, Sue Bastaja, Bev Benzakein, Anna Berezovskaya, Inna Berezovskaya, Lydia Block, Margaret Burton, Ruth Charney, Lucy Chu, Jack and Vivian Davis, Dana Frankel, Patricia Gallagher, Lynda Ginsburg, Eitan and Shaula Gurari, Karen Haylor, John and Barbara Hill, Katherine Hill, Fusa Hoshimura, Peggy Intrator, Ellen Kasulis, Gale Kingsley, Hutamu Kishimoto, Mary Beth Klotz, Sarah Kwon, Shirley Longshore, Patricia Lydon, Steven Reiss, Bill Schwartz, Hannah Schwartz, Jim Schwartz, Robert Schwartz, Cyndi Schultz, Robert Sickles, Carrie Strasburger, Jocelyn Tran, Ruth Ann Voit, Margit Winckler, Keith Yeates, and the librarians of The Ohio State University.

We would also like to express gratitude to our own teachers. Sadly, some of them are no longer with us, but the influence of them all lives on in the pages of this book: Pamela Dase, Tony DiLuna, Katherine Faber, Vicki Gartner, Robert Gurland, Marilyn Hoeflinger, Jerry Martin, Judith McDonald, Florence Mercer, and Dennis Readey.

Thanks go to Elizabeth Lindsmith and Lila Schwartz for their fine editing on earlier drafts and advice throughout the project; to Saskia Smith for translating some of the research; and to Terri Yoder—our Word whisperer, all-around assistant, and cheerleader—for guiding the manuscript and illustrations through their various configurations and revisions.

We are indebted to the Paul H. Brookes Publishing Co. Editorial Committee and our anonymous reviewers for their faith in this project. Special appreciation goes to our editor, Rebecca Lazo, and to Steve Plocher, Nicole Schmidl, Erin Geoghegan, and the rest of the Brookes staff for all their hard work in transforming our scruffy manuscript into a book.

Most of all, we wish to thank our families—our long-suffering husbands and children—for helping, encouraging, feeding, and humoring us, as needed, through this ambitious undertaking. We dedicate this book to you:

Bob   Jeff
Hannah, Lila, and Bill   Andrew and Matthew
(NK)   (SS)

Number Sense
and Number Nonsense

Chapter 1

Introduction

"Can you do Subtraction?" asked the Red Queen. "Take nine from eight."

"Nine from eight I can't, you know," Alice replied very readily: "but—"

"She can't do Subtraction," said the White Queen. "Can you do Division? Divide a loaf by a knife—what's the answer to that?"

"I suppose—" Alice was beginning, but the Red Queen answered for her.

"Bread-and-butter, of course. Try another Subtraction sum. Take a bone from a dog: what remains?"

Alice considered. "The bone wouldn't remain, of course, if I took it—and the dog wouldn't remain: it would come to bite me—and I'm sure *I* shouldn't remain!"

"Then you think nothing would remain?" said the Red Queen.

"I think that's the answer."

"Wrong, as usual," said the Red Queen: "the dog's temper would remain."

"But I don't see how—"

"Why, look here!" the Red Queen cried. "The dog would lose its temper, wouldn't it?"

"Perhaps it would," Alice replied cautiously.

"Then if the dog went away, its temper would remain!" the Queen exclaimed triumphantly.

Alice said, as gravely as she could, "They might go different ways." But she couldn't help thinking to herself "What dreadful nonsense we *are* talking!"

"She can't do sums a *bit*!" the Queens said together, with great emphasis.

From *Through the Looking-Glass*, by Lewis Carroll[1]

It is probably not a coincidence that the *Alice* books were written by a mathematics professor who delighted in children. Indeed, for many children, trying to learn mathematics is a lot like falling down the rabbit hole or stepping through the looking glass. In the Wonderland of the mathematics classroom as a child might see it, you meet really big numbers, like eight and eleven, and Tweedle Dee and Tweedle Dum lookalikes, like 6 and 9. Down the rabbit hole, you must do "five guzinta ten" (whatever that means) and add a and b (which are not even numbers)! On the other side of the looking glass, big numbers make fractions little, and little ones make them big. Sadly, despite your very best efforts, it is also where the Queen often stomps about declaring, "Wrong, as usual! She can't do sums a *bit!*"

It is sometimes easy to forget how utterly baffling even the simplest sum can seem to a child. Most children enter school with good mathematical intuition, but no one is born to the "dreadful nonsense" taught there. Our goal in this otherwise serious book for grown-ups is to take our readers inside young students' heads and show them mathematics through children's eyes.

Math difficulties have been studied deep in the shadow of research on reading disorder, their more prominent sibling in the learning disability family. Although recent technological advances have shed considerable light on certain aspects of mathematical cognition, readers seeking the level of enlightenment attained for dyslexia will be disappointed. The field is not ready to provide the kinds of popular books that now exist for reading disability, nor is it prepared to issue comprehensive curricular or pedagogical guidelines.

Nevertheless, science has by no means been silent on the subject of math learning; the purpose of this book is to gather, organize, and make accessible those discoveries. We approach this task from several different vantage points. First, to understand how learning goes wrong, one must understand how the mind "does" mathematics. From mastering counting to learning algebra rules, math draws on a vast range of mental abilities. On this topic, there has been copious research; however, studies are scattered in scores of journals across nearly a dozen academic disciplines. One major goal of *Number Sense and Number Nonsense* is to synthesize the current research on typical math cognition; because the pertinent findings are abundant and complex, that review is the focus of much of this book.

Unfortunately, mathematics does not come easily to everyone. It is vexingly abstract, is often counterintuitive, and relies on a dizzying array of quirky symbols, rules, and terms. Furthermore, people are equipped with different sets of cognitive strengths and weaknesses and prefer different ways of approaching problems (in life and in math). To comprehend why a child struggles with math, one must understand both the nature of mathematics and the mind of the individual student. The second goal of *Number Sense and Number Nonsense*, therefore, is to look at mathematics naïvely, as children do, and to examine what is known about individual differences in perception, cognition, and reasoning.

Teachers who possess insight into the cognitive demands that mathematical thought makes on children, as well as into the mysterious workings of their students' minds, will have a clear advantage in devising effective ways to explain the material. Although the scientific literature on pedagogy is slim, our third goal is to make that work available and link it to the broader findings.

Number Sense and Number Nonsense begins by briefly examining the history of the quest for insight into math learning difficulties. We also review the math disability definitions currently used in schools because a critique of some of the guidelines' basic assumptions will lead to the heart of this book, which is a scientific review of how the mind does mathematics and how that activity can often go awry.

A New Field

Scientists have been striving to unravel the mysteries of reading disability since the opening decades of the 20th century, but research on math learning difficulties has begun in earnest only recently. In 1937, British psychologist Erich Guttmann was the first to write about "congenital" math learning difficulties.[2] Following that initial paper, research at first proceeded very slowly. Then, in 1977, the U.S. Office of Education set forth the first legal definition of specific learning disabilities; 3 years later, the American Psychiatric Association radically updated its inventory of mental disorders and included a Developmental Arithmetic Disorder.[3] Since that time, research into both typical mathematical cognition and math learning difficulties has accelerated rapidly. Such diverse fields as animal behavior, infant and early childhood cognitive development, education, medicine, linguistics, genetics, experimental and cognitive psychology, neuropsychology, and—most remarkably—structural and functional brain imaging have all contributed to our knowledge of mathematical cognition. In 1985, McCloskey and his colleagues put forth the first comprehensive model of normal mathematical cognition, which still serves as the foundation for much mathematical disability research.[4]

Why have researchers been so much slower in tackling math learning difficulties than in dealing with dyslexia? Do fewer children have trouble with math than with reading? Are teachers more willing to accept weakness in math? Are math difficulties less debilitating in school or in life? Do general classroom teachers feel less confident teaching math and thus pay less attention to children who struggle with it? And why is it that children with isolated math difficulties so infrequently receive a referral for diagnostic evaluation or assignment to special education?[5] These questions have puzzled scholars for a long time. As Guttmann wrote in 1937,

> The number of cases referred to the children's department of [our clinic] as backward in arithmetic is surprisingly small, when compared with the cases of reading disability. In a period in which we were able to collect 50 cases of backward readers, we observed only 6 or 7 cases of arithmetic disability. It is debatable whether this is due to the subject matter or to the different attitude of the educationalists toward it.[6]

Perhaps the recent surge in research and the insights it provides will spark an interest in identifying the children who need help and in finding effective ways to do so.

What Is a Mathematics Disorder?

Since the late 1970s, psychiatric and governmental guidelines have determined how students are classified as "learning disabled." The legal definition, in particular, has determined eligibility for special education and for this reason carries considerable educational and political weight. The next two paragraphs introduce these definitions; we discuss them in greater detail in the following sections.

The current authoritative working definitions of mathematics disorder come principally from two sources. The first definition, from the *Diagnostic and Statistical Manual of Mental Disorders, Fourth Edition, Text Revision (DSM-IV-TR)*, identifies Mathematics Disorder as calculation or math reasoning ability falling significantly below the level expected on the basis of age and measured intelligence, assuming that the student has received adequate education. The disorder must be debilitating in school or in life and must not be solely due to a vision or hearing deficit. The guidelines note that emotional disorders and impairments in language, perception, and attention are frequently associated with Mathematics Disorder, which is often evident early in the course of mathematics education.[7]

The second, legal definition comes from the Individuals with Disabilities Education Improvement Act of 2004 (IDEA 2004; PL 108-446), which describes a specific mathematics learning disability as "a disorder in one or more of the basic psychological processes involved in understanding or in using language, spoken or written, that may manifest itself in the imperfect ability to . . . do mathematical calculations." This definition includes conditions such as perceptual, reading, and language disabilities, as well as math impairments acquired through brain injury. Specifically, a child is determined to have a math disability if he or she "exhibits a pattern of strengths and weaknesses in performance, achievement, or both," in mathematics problem solving or calculation "relative to age, State-approved grade-level standards, or intellectual development," when provided with age- and grade-appropriate learning experiences and instruction, and as demonstrated by serial formal skills assessments. Like the *DSM-IV-TR* guidelines, IDEA 2004 excludes primary sensory, motor, emotional, cultural, and other environmental causes, as well as math difficulties primarily due to intellectual disabilities. The IDEA 2004 definition eliminates a previous legal requirement for a statistically formulated discrepancy between achievement and intelligence and instead leaves the diagnosis to professional judgment. Finally, it allows for a diagnostic determination based either on standard diagnostic testing or on the child's failure to respond to "scientific, research-based intervention."[8]

Both the psychiatric *(DSM-IV-TR)* and the legal (IDEA 2004) definitions of mathematics disability raise several important issues. The following sections address some of the definitions' assumptions and the difficulties inherent in defining and studying the problem of math disability.

Is There Such a Thing as a Math Disorder?

To determine if there is such a thing as a math disorder, one must look to behavioral genetics. In general, this field explores genetic and environmental contributions to individual variation in different traits by studying differences in twins' behavior. For example, by examining the math achievement of large groups of identical twins (who share 100% of their genes) and fraternal twins (who share only 50% of their genes), scientists discovered that about 50%–70% of differences in math achievement are influenced by genetics. However, this does not necessarily mean that our genes limit our math abilities. Education, family, and cultural environment, as well as many other factors, interact with our genetic endowment in myriad complex ways to make each individual unique. A person's math ability reflects all of those influences.[9]

Behavioral geneticists also found that the genetic contribution to math *disability* was of the same magnitude as that to general math *ability*. Molecular geneticists have not yet discovered the genes that affect math learning, but they expect that there will be many such genes, each with a very small influence on many different brain mechanisms. The behavioral findings suggest that the genes influencing math disability will be the same ones that influence the full range of math skill. Researchers therefore concluded that math disability is not an all-or-none disorder. No qualitative difference exists between students with severe difficulty and students with typical math abilities—a conclusion similar to one reached regarding reading and language ability and disability.[10]

There are, of course, quantitative differences among different students' math skill. It does not take scientific research to tell us that some people are better at math than others! Like most other human traits, math learning ability ranges from very strong to very weak, as described by a "normal" (bell) curve, with most people falling in the middle. That curve is smooth, with no natural demarcation between those with and without a learning disability.[11] Any line separating those categories is not rooted in science; rather, it is arbitrary and subject to conflicting definitions and rationales.

It is possible to develop a statistical definition of math disability that is reliably tied to measurable debility—what psychologists call *predictive validity*. Medical diagnosis often makes use of this method; hypertension, for example, is defined as the part of the blood-pressure range most reliably linked to heart disease and stroke. One study demonstrated that young adults' daily-life arithmetic competence predicted the probability of full-time employment (above and beyond reading and writing skills).[12] Using the math cutoff developed from this study, one can then roughly determine who is at risk for unemployment. Similarly, scientists are now looking for ways to identify very young children who will be at risk for math failure in elementary school.[13] Caution is warranted, however: This method of identification is also arbitrary to the extent that choosing a predictive criterion (e.g., full-time employment, third-grade math failure) is a matter of judgment.

The problem of definition affects disability research and educational practice.[14] The researchers cited in this book who compare "disabled" and "nondisabled" groups usually do so on the basis of some cutoff defining math disability (e.g., the bottom 5%, 10%, or 25% on a standardized math test). This convention is not necessarily unreasonable; however, it is important to remember that the cutoffs are simply selected to suit the various studies, not because there is some scientifically determined qualitative difference between children on either side of the divide. In addition, two studies with different severity criteria may produce incomparable results. In the laboratory, researchers can avoid this problem by examining the links between math learning and related factors across the whole ability range; in fact, many (but certainly not all) of the studies reviewed in this book have approached the problem in this manner.

It is also important to remember that measurement is never exact and is always subject to error. Measurement of achievement (in math or any other skill) is rarely more than rough accounting; as such, it provides an unreliable estimate of an individual's true abilities. Psychometric crudeness makes it difficult to rationalize that a test score slightly more or less than the cutoff is truly different from a score just to the other side. For example, on a test for which the 20th percentile is the chosen cutoff, one cannot justify that a student scoring at the 18th percentile has a math disability, whereas one performing at the 22nd percentile does not. This is not very different from the common situation in which a student regularly fluctuates between a *B+* and an *A−*, often depending on the nature of the test. One would have difficulty identifying that student as an *A* student or a *B* student. Cutoffs are not only arbitrary, they are also blurry.

Despite the abundant difficulties inherent in defining a discrete math learning disorder, there is a legitimate political and economic reason for doing so: to allocate educational resources. In the schools, however, drawing an artificial line between children who do and do not qualify for special educational services—although convenient and presumably objective—may deny services to children who do not meet the diagnostic criteria but still need help. As anyone who has been involved in such a situation knows, these decisions can create an adversarial mood within a school community and can often be educationally unsound. The evaluation format recommended by IDEA 2004 includes multiple other measures, such as classroom observation and repeated math assessment, which may help to reduce the chance of overlooking a student in need of help.[15] It is important for teachers, parents, and students to understand that the official cutoff criteria for math disability (or any other specific learning disability, for that matter) are based more on political and economic considerations than on science.

What Is Intelligence and How Is It Measured?

The psychiatric diagnostic guidelines in the *DSM-IV-TR* define math disability in part by the discrepancy between achievement and "measured intelligence," a strategy that has come under scholarly fire in recent years. For several decades, the legal definition was based on a statistical formula defining a severe discrepancy between math achievement and overall intelligence. IDEA 2004 eliminated this requirement, although it retained an (unelaborated) option of comparing achievement or classroom behavior to intellectual development.[16] Including a measure of intelligence in the definition allows students whose math performance meets grade expectations but falls short compared with their overall abilities to be considered "disabled" and thus eligible for services. Some people believe that entitling such students to special education is a poor use of scarce resources, whereas others feel that those are the very students who could make the best use of extra educational assistance. Advocates on both sides of the debate cite the pressure of global economic competition, among other rationales, to support their views. Meanwhile, many students with strong intellectual capabilities have felt very frustrated by their puzzling, isolated difficulties in mathematics.

First, it is important to determine what is meant by *intelligence*. Scientists have debated the meaning of intelligence for more than a century. Some argue that there is such a thing as a core measurable intelligence, whereas others assert that intelligence is merely a composite of one's disparate mental abilities—an average of all of one's strengths and weaknesses. For the most part, psychologists recognize that both positions are valid and have defined intelligence as follows:

> Intelligence is a very general mental capability that, among other things, involves the ability to reason, plan, solve problems, think abstractly, comprehend complex ideas, learn quickly and learn from experience. It is not merely book learning, a narrow academic skill, or test-taking smarts. Rather, it reflects a broader and deeper capability for comprehending our surroundings—"catching on," "making sense" of things, or "figuring out" what to do.[17]

On the one hand, IQ test analyses have shown that some kinds of abilities, especially reasoning, are closely linked to all the others and seem to reflect an abiding mental capacity that could be called one's intelligence. This conclusion is supported by behavioral genetic studies, which have found that various mental skills are genetically linked to one another to some degree. Like achievement, intelligence is characterized by a strong-to-weak bell curve and thus also lacks natural cutoffs.[18]

On the other hand, clinical studies have shown that although some people's abilities are relatively consistent with one another, most people's cognitive landscapes are not completely flat. These landscapes consist instead of rolling hills, or in some cases even peaks and valleys—the diverse strengths and weaknesses that make an individual unique.[19] Someone who is strong in mechanical reasoning may be weak in verbal logic, for example. In addition, brain injury can often affect some cognitive skills while leaving others untouched, suggesting that one's mental fabric is not seamless and that marked cognitive variability within an individual is at least neurologically possible. Indeed, substantial disparity often characterizes individuals with specific academic difficulties. Therefore, using measurable intelligence in the diagnosis of math disability is problematic when a single number is assigned to represent a disparate array of abilities. For most people, a summary IQ score is much less meaningful than a set of measures that describes their range of cognitive strengths and weaknesses, a diagnostic strategy absent from the disability identification procedures delineated in IDEA 2004.[20]

A second difficulty in defining math achievement against intelligence is that math ability is not independent of cognitive abilities; indeed, it is to some extent a product of them. In fact, behavioral genetics has demonstrated that about one third of the genetic contribution to math ability differences is tied to differences in general intelligence.[21] This partial link between intelligence and achievement complicates measuring intelligence apart from math-related mental skills; many children may have significant difficulty with math while performing in a manner consistent with their measured intelligence. As noted, IDEA 2004 does not recognize a distinct math disability in students with intellectual impairments. Clinical studies have linked intellectual disabilities to both severity and persistence of math disability.[22] Among primary-grade children, a discrepancy between achievement and IQ did not provide useful distinctions for predicting the severity of disability, risk of poor achievement, or performance levels on math word problems or working memory tests, throwing its practical value into question. These findings have prompted some researchers to suggest that general intelligence should be taken out of the diagnostic equation altogether.[23]

The genetic link between math ability and intelligence, however, is only partial: Doing well in math only partly depends on how intelligent an individual is. In fact, about one third of the genetic influence on math performance is specific to math apart from both overall intelligence and other academic skills. Many other factors play a role in developing math skills. As a result, some students may perform more poorly in math than they do in other academic areas, such as reading comprehension, or on tests of reasoning and other measures of intelligence, regardless of their ability range in any of these areas. Thus, scientific evidence supports the common-sense notion that a performance discrepancy—where it exists—between math and other academic skills or between math and intelligence may signal worrisome difficulties in math learning. Even though the statistical criteria by which one would determine such a discrepancy are problematic for all of the reasons previously noted, measures of intelligence and other academic skills should be an essential part of these children's evaluations. IDEA 2004 encourages assessing achievement patterns against intellectual development, and educators are searching for a valid and reliable strategy that would allow them to incorporate such information into eligibility considerations while avoiding the pitfalls of a severe-discrepancy formula. Policy makers in most states are wrestling with this perplexing conundrum.[24]

A third difficulty in relying on measurable intelligence is that not all forms of intelligence relevant to math are easy to measure. Intelligence experts have posited a kind of "general-purpose associative learning system"—a tacit, pragmatic kind of street smarts. This kind of intelligence has not been captured or quantified by IQ tests; nevertheless, it figures heavily in the specific, job-related math competencies of many children and adults who are either unschooled or did not do well in school. Studies have revealed that many unschooled individuals from various walks of life have an astounding ability to master math tasks directly related to concrete goals. When asked to perform the same tasks abstractly using pencil and paper, however, they are often unable to do so. Further, most of them did not perform well on either unrelated math tasks or on standard IQ tests. (Notably, educated novices performed no better, and sometimes worse, than unschooled experts on the practical tasks in these studies.) These individuals learn what they need to know implicitly through experience—often driven by the will to succeed, the fear of potentially catastrophic failure, or the desire to get the job done and go home early.[25] These practical math skills are important to consider, although they may not show up in conventional measurements of math ability.

Prevalence and Persistence

Defining a math disorder by arbitrary cutoffs creates statistical problems that make it difficult to identify its prevalence and persistence in the general population. What proportion of the student population has a math disorder? Teachers and district superintendents require this sort of information to determine classroom and general personnel needs. Relying only on a math test cutoff to determine prevalence, however, quickly devolves into circular reasoning. Given a large enough selection of children, the proportion scoring below the 10th percentile (a commonly used cutoff) on a standardized test will be, by definition, 10%. If one excludes students whose math difficulties are best explained by social or cultural factors, the prevalence will be somewhat less. And, indeed, scientists have conservatively estimated the prevalence of math disorder in the general population at 3%–8%, but, of course, varying the criteria and statistical techniques used to analyze the data has produced different estimates.[26] The published estimates of math disability prevalence among students with general learning disabilities—50% among younger individuals to 60% or more among older students and adults—also depend on working definitions with arbitrary cutoffs.[27]

In addition, because of measurement variability, students' performance on a math test taken one day may differ from their performance at a later date, even if their abilities have not significantly changed in relation to those of their age mates. This measurement instability can make it difficult to assign students to a diagnostic category or to gauge the disability's persistence, especially if those scores wobble back and forth across the diagnostic divide.[28] Knowledge of persistence comes chiefly from a handful of longitudinal studies confirming, very broadly, that only about half of the children who demonstrated math disability at one point continue to satisfy diagnostic criteria at a later date. About one third of genetic influences vary by age; however, the factors influencing growth and change in an academic subject such as math tend not to be related to genes but rather to idiosyncratic personal characteristics and, of course, education. In these longitudinal studies, the pattern of 50% persistence remained true regardless of the age of first math disability diagnosis (i.e., kindergarten or grades 4–7), the presence of reading or spelling disabilities, the amount of time between assessments (i.e., 19 months to 6 years), the diagnostic test cutoff, or (discouragingly) special education interventions.[29]

So far, no general scientific agreement has been reached about what differentiates students who move out of the disability category from those who do not, although researchers are now attempting to develop early childhood screening instruments that can effectively predict persistent math trouble in the early grades. Because a single assessment may or may not signal a lasting disability, researchers recommend that consistently poor performance over several years be incorporated into any definition of the disorder to compensate for both measurement error and the normal ebb and flow of development and learning. As discussed, federal regulations in IDEA 2004 require serial assessments of classroom performance, although they do not specify over what time period. Of course, determining the best duration for longitudinal assessment is also problematic: If it is too short, one risks overdiagnosis; if it is too long, one risks missing an opportunity for early intervention.[30]

What Kind of Math Difficulty Counts as a Disorder?

Diagnostic guidelines for a math disorder include consideration of difficulties in both calculations and problem solving, but they are not very clear about how best to measure those skills, particularly at different age and grade levels. Just as researchers have chosen various severity cutoff scores for determining math disability, they have also chosen different ways of measuring math abilities. Several studies (especially the older ones) used an untimed, mixed-operation test of fact retrieval or simple mixed calculations. Other studies used some combination of tasks, including calculations, oral or written word problems, estimation, or for very young children, a combination of informal and early conventional skills. Some researchers have based their tests on theory, whereas others have relied on the curriculum. Still others do not specify their rationale. This use of different math tasks to determine disability makes it difficult to compare results across studies.[31]

Behavioral genetics research has revealed that, on average, people who find some math tasks hard will find most others hard as well.[32] However, many other factors besides genetics make some types of math problems more difficult than others for some students. Many of those influences are discussed in the ensuing chapters. In the meantime, it is important to understand that there is no clear consensus regarding precisely what kind of math difficulty defines a disability.

What Is Age-Appropriate Math Education?

Both legal and psychiatric diagnostic guidelines in IDEA 2004 and the *DSM-IV-TR*, respectively, correctly note the importance of assessing the quality of a child's math education when determining whether or not a child has a learning disability. Math education begins well before children start formal instruction in the primary grades. Informal experiences in the home, on the playground, and at preschool play a significant role in establishing the most fundamental math skills. In fact, there is clear evidence that children who enter kindergarten from homes and preschools rich in informal math experience, such as counting and playing certain types of board games, have a significant head start over children lacking those early benefits.[33] (Notably, formal elementary math education often inadvertently abandons informal concepts; in gaining technical skills, children frequently lose their intuitive ones.[34]) Later in this book, we examine some recent research into determining early risk factors for later math learning difficulties, as well as basic activities that can ameliorate those early deficiencies and keep the early intuitions alive.

Although the ingredients of age-appropriate informal math education are becoming clearer, little is yet known about what constitutes age-appropriate formal math education. As the National Mathematics Advisory Panel noted in their 2008 report, no scientific data yet support one curriculum over another. Moreover, there has been little scientific evaluation of mathematical pedagogy.[35] Educators still hotly debate the correct balance between explicit instruction and "constructivist" experimentation, for example. Even the proper method for teaching and drilling arithmetic facts remains an open question. These issues are difficult to study because many factors affect children's classroom experience—for instance, curriculum; their teachers' experience, motivation, knowledge, philosophy, and personal style; the theoretical bent of textbooks; the intellectual, personality, and demographic mix of students in the classroom; and administrative and community support. Sorting through each factor's effect and cumulative influence on student comprehension is a mammoth undertaking. Moreover, for any given student, the fit between the instructional approach and the child's individual cognitive strengths and weaknesses is largely unexplored territory. In short, determining whether a student has had the benefit of age-appropriate education is far from simple.

In a related matter, some experts have recommended including the child's response to intervention in the learning disability diagnostic protocol, an evaluation method permitted by IDEA 2004. In particular, IDEA 2004 specifies that the diagnostic intervention be "scientific [and] research based." However, response to intervention assumes that a body of knowledge exists regarding effective pedagogy. Research-based instructional methods are available for reading; for children with suspected dyslexia, one can identify age-appropriate education and apply effective diagnostic interventions.[36] In mathematics, however, only a few studies have delineated some techniques that have reliable diagnostic outcomes at certain ages. One comprehensive study, for example, found that 16 weeks of adjunctive tutoring provided modest diagnostic discrimination among a group of at-risk first graders, depending on the measures and criteria employed.[37] Research has a long way to go before it produces scientifically based math instruction sufficient to use for diagnostic purposes or in the general classroom.

The Focus of this Book

So far, we discussed the problems inherent in categorizing students who struggle with math. Finding a reliable method for doing so is critical for administrators and public policy officials, who must make decisions affecting not only large numbers of students, but also the allocation of personnel and other economic resources. For some of these professionals, individual student differences are noise in the system, creating problems for sorting students into class placements or services. By contrast, teachers and psychologists (including this book's two authors and probably many of its readers) regard students' individual differences as a fact of everyday classroom life. For them, it is less important to know what category a child fits into than what makes math difficult for that particular child—a conundrum that goes to the heart of teaching. Indeed, the National Council of Teachers of Mathematics noted that the best teaching will take into account students' particular learning needs.[38]

In writing this book, therefore, we chose to step away from the question of what defines a math learning disorder. Instead, we focus on these important individual differences. Specifically, we address general psychological and educational issues: the cognitive demands of mathematics, the cognitive resources of the students, and—when possible—the pedagogy required to link the two.

What cognitive demands does learning mathematics place on a student—*any* student? For example, when a student is asked to perform a complex calculation or solve a word problem, what mental resources will the student need? To what extent are these skills general or particular to math? What can we learn about typical mathematical thought processes from people with known cognitive disabilities? How does this information help us make sense of the difficulties that both typical students and those with disabilities encounter with math? As it turns out, these questions are exceptionally complex.

An early study of two children, one "gifted in arithmetic and another with an arithmetic defect," concluded that "neither in the abnormally good nor in the abnormally bad case were effects limited to calculating," casting doubt on the "specificity of defect and gift." The researchers felt "unable to decide with which psychic capacities arithmetic gift and arithmetic defect are correlated."[39]

Since that study, the quest to identify which cognitive abilities are essential for math has intensified, but findings have not yet coalesced into a clear picture. The body of research literature on math ability and disability is like the proverbial group of blindfolded people straining to describe an elephant: Some say it is like a leather sail, others like a rope, and yet others like a garden hose or a tree trunk. Some scholars insist that the defining characteristics of math disability must be specific to math alone, or else math ability would simply be synonymous with intelligence.[40] Other researchers have arrived at a single mental characteristic—although this characteristic differs across studies.[41] Diverse research has linked math achievement to spatial skills, working memory, phonological processing, cognitive flexibility, and more.[42] Some studies—noting that there may be several underlying core impairments—have described math disability subtype schemes based on various factors, including verbal versus spatial skills or math error patterns, whereas other research has tried to distinguish between math difficulties with reading disorders and those without them.[43] The few large-scale studies that have fished for patterns of underlying neuropsychological impairments linked to math failure have been inconclusive.[44] Ultimately, there are few cognitive abilities that have not been associated with math achievement—and few cognitive impairments not linked to math failure.

Overall, the research suggests that if several specific cognitive weaknesses are linked to math difficulties, the weaknesses may be somewhat different for different children. Indeed, no logical reason exists to assume that there is a single, definitive math disorder; the evidence, taken as a whole, points to an array of potential underlying difficulties. As we discuss in the next few chapters, many perceptual, cognitive, executive, and reasoning skills are necessary for math. It is not unreasonable to speculate that an impairment in any of those areas may make math especially difficult, particularly if teaching methods do not acknowledge and address those weaknesses. Given this state of affairs, diagnosing a disorder is less useful than describing a child's math troubles in relation to that child's individual pattern of perceptual, cognitive, executive, and higher reasoning strengths and weaknesses.

In walking our readers through the maze of research on math ability and disability, we are broadly inclusive as we report on findings related to a wide range of other difficulties including language disorders, dyslexia, and ADHD—the random mix of issues students bring to the classroom. Additionally, we include math difficulties brought on by neurological injury, chiefly to illustrate brain function. In another sense, however, we focus narrowly by stressing the perceptual, cognitive, executive, and higher reasoning factors in math learning. We are mindful of the many social and emotional contributions to learning, including personality, motivation, outlook, self-confidence, and home and classroom climate, which we refer to occasionally—but only briefly. We decided to restrict our discussion in this manner in part because we had to limit the scope of the book somehow, and those topics deserve a volume of their own.[45] Although research has arrived at some generalizations about these factors' powerful effects on math learning, how they play out for any given child is complicated and less predictable. Moreover, for a student with a specific learning impairment and/or inappropriate instruction, anxiety and poor motivation may simply be rational responses, albeit complicating ones. Intellectual and emotional factors interact in highly complex ways, however, so we try to avoid simplistic distinctions.

Number Sense and Number Nonsense also focuses on the professional implications of the scientific research. One of our goals is to help the teacher and psychologist get into the mind of the student. If one understands a student's cognitive profile and the cognitive skills that math demands, then one ought to be able to design a teaching approach that links the one with the other. Appropriate pedagogy is important, not just to reduce students' frustration, but because learning develops the brain. How one is taught affects how one grows, and all abilities must be considered work in progress. In assessing pedagogical techniques, one should ask: For whom is a certain method effective? How does this student understand this material, and what is required to clear the roadblocks? Drawing the educational implications of the research was our toughest challenge as authors, because there has been very little scientific study of mathematical pedagogy for typical students, much less for those who struggle with math. We offer our findings and ideas with the hope that what we have discovered in the scientific literature to date might be useful to those who shape math pedagogy.

Chapters 2–9 form the core of *Number Sense and Number Nonsense*. In them, we review the scientific literature on mathematical cognition. These chapters take their organizational cue largely from what we know about the functional organization of the brain.[46] Chapters 2 and 3 focus on the informal, unschooled sense of quantity—notions that are closely linked to spatial awareness. On close reading, one sees that the legal definition of math disability considers only linguistic impairments, as we have seen, but research indicates that nonverbal skills also play a significant role in mathematical thought. Psychological and neuroscientific findings suggest that one thinks of numbers as if they were lined up from small to big, as on a ruler. These chapters examine how students develop this mental number line and, more broadly, focus on the spatial, dynamic, and mechanical reasoning skills necessary for math and its many applications. Chapter 3 also explores how natural imaging preferences affect math ability and speculates about the silent role that spatial reasoning impairments may play in math difficulties. Finally, these chapters address two classroom challenges facing teachers: Can number sense and spatial skills be taught?

Chapters 4–7 examine the mathematical language, including counting and conventional notation, that forms the heart of traditional math education. Although some societies rely on sticks, stones, and body parts for accounting purposes, the mathematics of modern civilization depends upon spoken and written mathematical language. These chapters focus on the relation between natural and mathematical language, the linguistic skills necessary to master conventional mathematics, the math difficulties resulting from perceptual and language impairments, and the relationship between reading and math disabilities. They examine the development of counting and simple arithmetic, as well as the linguistic demands of such vexing tasks as complex computations, word problems, fractions, and learning math facts and terms. They also explore the issue of bilingual math education and review neuroscientific findings that elucidate the way students do formal mathematics. Finally, they look at multisensory classroom techniques borrowed from reading instruction and at popular but untested nonverbal aids to teaching conventional math.

Chapters 8 and 9 delve into general mental abilities—the so-called executive functions and higher reasoning skills—that are not specific to mathematics but are nevertheless vital for it: attention, mental control, working memory, planning and sequencing, self-monitoring, mental flexibility, abstract reasoning, logical thought, and hypothesis testing. These chapters discuss how these functions develop, how they relate to mathematical problem solving, and how teachers can foster them in the classroom. These chapters also explore math difficulties often found in ADHD and other conditions, such as anxiety, as well as pedagogical techniques designed to help students with these conditions.

The final two chapters directly address professional issues. Chapter 10 draws on what is known about the abilities necessary for math to suggest strategies for evaluating the student's difficulty: what kinds of questions to ask students and their parents, how to assess math performance, and how to test underlying cognitive abilities. The chapter offers guidelines for both the classroom teacher and the evaluating psychologist. Chapter 11 focuses on the importance of thinking about mathematical education in terms of both the cognitive demands of the discipline and the cognitive strengths and weaknesses that each student brings to it in a standards-based educational system. In so doing, the chapter summarizes the classroom ideas presented in the research chapters and draws implications for classroom assessment options. It also briefly reviews various pedagogical theories and controversies, including the matter of teacher preparation, and suggests ideas for future research that will lead to solid pedagogical recommendations. Finally, it ties the scientific findings to the recommendations of the 2008 National Math Advisory Panel.[47]

Throughout *Number Sense and Number Nonsense*, we provide case illustrations, taken from both the classroom and the psychologist's office, of students whose battles with mathematics have felt hopeless. The cases demonstrate the complex relationship between one's cognitive profile and mathematical ability, while conveying the students' frustration and the urgent need to find help for them.

Number Sense and Number Nonsense is the first attempt to synthesize the vast body of research on mathematical cognition; in that sense, it ventures into uncharted seas. Far from being the last word on the many issues raised herein, it is one of the very first words; as such, it is designed to encourage scholars in cognitive development, neuroscience, cognition, and educational psychology to move forward with research that addresses the many questions that remain unanswered. The book also aims to provide a child's-eye view of mathematics for the classroom teacher and the school psychologist. We hope that whatever insights professionals gain will enable them to find new ways to reach not only the struggling student who "can't do sums a *bit!*" but also those typical students who sometimes, well, just don't get it.

Notes

[1]Carroll, 1872/1960b, pp. 320–321.
[2]Guttmann, 1937.
[3]U.S. Office of Education, 1977; American Psychiatric Association, 1980.
[4]McCloskey, Caramazza, & Basili, 1985.
[5]See McLeod & Armstrong, 1982.

[6]Guttmann, 1937.

[7]American Psychiatric Association, 2000.

[8]IDEA 2004 Final Rule, 2006, §300.8(10), §300.307, §300.309. See also Lichtenstein & Klotz, 2007.

[9]Plomin, Kovas, & Haworth, 2007; Shalev et al., 2001.

[10]Plomin et al., 2007; Shaywitz, Escobar, Shaywitz, Fletcher, & Makuch, 1992.

[11]Plomin et al., 2007.

[12]Rivera-Batiz, 1992.

[13]See, e.g., Murphy, Mazzocco, Hanich, & Early, 2007.

[14]For issues addressed in the ensuing discussion, see Francis et al., 2005; Murphy et al., 2007.

[15]IDEA 2004 Final Rule, 2006, §300.306, §300.309.

[16]IDEA 2004 Final Rule, 2006, §300.307, §300.309. See also Francis et al., 2005; Lichtenstein & Klotz, 2007.

[17]Gottfredson, 1997, p. 13.

[18]Mackintosh, 1998; Petrill, 2002; Plomin et al., 2007.

[19]See, e.g., Wechsler, 2003, Appendix B, Table B.6, p. 266; Wechsler, 1997, Appendix B, Table B.5, p. 211.

[20]See Lichtenstein & Klotz, 2007.

[21]Plomin et al., 2007.

[22]Murphy et al., 2007; Shalev, Manor, & Gross-Tsur, 2005.

[23]Gonzalez & Espinel, 1999; Mazzocco & Myers, 2003.

[24]Ahearn, 2008; IDEA 2004 Final Rule, 2006, §300.309; Lichtenstein & Klotz, 2007.

[25]Mackintosh, 1998, pp. 360–363. See also, e.g., Schliemann & Nunes, 1990; Ginsburg, Posner, & Russell, 1981.

[26]Barbaresi, Katusic, Colligan, Weaver, & Jacobson, 2005; Desoete, Roeyers, & DeClercq, 2004; Geary, 2003; Shalev, Auerbach, Manor, & Gross-Tsur, 2000.

[27]Shalev et al., 2000; Bryant, Bryant, & Hammill, 2000; U.S. Department of Education, National Center for Education Statistics, 1997, pp. 46–47.

[28]Francis et al., 2005.

[29]Kovas, Haworth, Dale, & Plomin, 2007; Mazzocco & Myers, 2003; Shalev et al., 2005; Silver, Pennett, Black, Fair, & Balise, 1999.

[30]See Geary, 2003.

[31]See, e.g., Badian, 1983; Gross-Tsur, Manor, & Shalev, 1996; Landerl, Bevan, & Butterworth, 2004; Mazzocco & Myers, 2003; Rourke, 1993.

[32]Kovas, Petrill, & Plomin, 2007.

[33]Ramani & Siegler, 2008.

[34]See, e.g., Jordan, Hanich, & Kaplan, 2003a.

[35]National Mathematics Advisory Panel, 2008.

[36]Shaywitz et al., 2004.

[37]Fuchs et al., 2005.

[38]National Council of Teachers of Mathematics, 2000.

[39]Lazar & Peters, 1915, as cited in Guttmann, 1937, p. 16.

[40]Chiappe, 2005.

[41]See, e.g., Iuculano, Tang, Hall, & Butterworth, 2008; Swanson & Jerman, 2006.

[42]In addition to the studies cited throughout Chapters 2–9, see Floyd, Evans, & McGrew, 2003.

[43]See, e.g., Geary, 1993; Jordan, Levine, & Huttenlocher, 1995; McCue, Goldstein, Shelly, & Katz, 1986; Robinson, Menchetti, & Torgesen, 2002; Rourke, 1993; Von Aster, 2000.

[44]See, e.g., Clarren, Martin, & Townes, 1993; Hale, Fiorello, Bertin, & Sherman, 2003; Rosenberger, 1989; Shalev, Manor, Amir, Wertman-Elad, & Gross-Tsur, 1995; Shalev, Manor, & Gross-Tsur, 1997.

[45]For readers interested in the work in these areas, see, e.g., Blackwell, Trzesniewski, & Dweck, 2007; Graham & Golan, 1991; Guttman, 2006; Linnebrink, 2005; Meece, Anderman, & Anderman, 2006; Pajares, 2005.

[46]See Reitan & Wolfson, 2004, for a conceptual framework based on brain function.

[47]National Mathematics Advisory Panel, 2008.

Thinking Spatially

When most people think of mathematics, they think about arithmetic facts, equations, proofs, and the like. This math requires an ordered list of counting words and matching written numerals. It also requires other language that tells us what to do with the numbers, such as *multiply* or *take the square root*, or that describes how shapes relate to each other, such as *parallel* or *congruent*. Furthermore, it includes a long list of rules that explain how they all work together. This is the math we study in school.

But what if mathematical language and symbols did not exist? Could people think about quantity at all? Section I examines the intuitive side of mathematics and how that intuition influences people's grasp of the concepts that the conventional symbols represent. The old philosophers appreciated that mathematical intuition is intimately tied to one's sense of space and time; by envisioning objects and events in those dimensions, one comes to understand the structure of our universe, the stuff of mathematics.[1] These mental pictures help people not only to understand formal mathematics, but also to gain new mathematical insights. As Albert Einstein famously remarked,

> Words and language, whether written or spoken, do not seem to play any part in my thought processes. The psychological entities that serve as building blocks for my thought are certain signs or images, more or less clear, that I can reproduce and recombine at will.[2]

Chapter 2 explores how spatial insight affects one's understanding of number and quantitative problem solving. Chapter 3 shows how it informs geometry, map and model reading, and proportional and mechanical reasoning—significant branches and applications of mathematics. In the course of that discussion, we also examine how failing to form spatial images can compromise math achievement for some students.

Notes

[1] See Boroditsky, 2000.
[2] As cited in Dehaene, 1997, p. 151.

Chapter 2

Number Sense

Science has demonstrated that humans are not alone in the ability to quantify. Animals, which do not have "words and language, whether written or spoken," use numbers every day for survival.[1] Across the phylogenetic spectrum from insects to chimpanzees, survival of both the individual and the species depends on quantitative skills to communicate, forage, evaluate threat, track offspring, optimize breeding, and conserve energy. Animals' remarkable quantitative abilities include a sense of *how many* (e.g., eggs in the nest) and *how much* (e.g., distance from predators); some animals have even been trained to determine *which one* in a series (e.g., the third tunnel in a rat's maze). Animals make these judgments based on information obtained through all of their senses.[2]

Even though they cannot yet talk, human infants also have a rudimentary sense of quantity. For example, infants can distinguish two cookies from three and know that adding to, or taking away from, a small number of toys brings predictable results. In studying infants' mathematical skills, scholars debate two key issues. The first is whether these very young children can clearly distinguish between *how many* and *how much*. For example, in judging whether two cookies are the same as three cookies, they may be basing their decision on the amount of "cookie stuff" (i.e., the surface area or volume of the cookies) or on the cookie numerosity (i.e., the countable number of cookies). (A note about terminology: In this book, we use various terms for number. *Numerosity* refers to the amount of discrete things or events in a collection and is a property of the collection itself. *Numeral* is the written symbol. We use *number* generically, in the singular or plural, to denote discrete quantity regardless of format.)

Unfortunately, infants are too young to tell us what they are thinking. So far, research has been inconclusive and the debate remains lively. For some scientists, the resolution of this first issue hinges on the second key question: whether human quantitative abilities are directly inherited from animals through evolution or rather represent a specifically human adaptation of more basic shared perceptual skills. Scholars continue to debate this thorny question as well.[3]

The Mental Number Line

Researchers do agree, however, that both animals and young humans have a rudimentary sense of quantity. These primitive quantitative notions have two striking qualities: They are relative and approximate. Without being able to count, both animals and human infants judge quantities in relation to other quantities. For example, two cookies take on quantitative meaning only in relation to one cookie (i.e., more) or three cookies (i.e., less). Many animals live or die based on their skill at determining *more*—pertaining not just to continuous amounts such as water, distance, or time, but also to discrete amounts of food bits, predators, and eggs. Indeed, many creatures can weigh one factor against another in a truly remarkable cost-benefit analysis.

Young children first demonstrate a rudimentary, implicit sense of *more* around the end of their first year or early in their second year, when they pick three cookies over two. During their third year, when they can purposefully manipulate objects and understand instructions (but before they can count reliably), children make explicit ordinal judgments (e.g., picking the "winner," or larger, of two small collections of boxes). By age 5 years, children can compare numerosities from memory, suggesting they have a mental representation or image of them.[4]

Because humans' early idea of numerosity is relative and therefore ordered, scholars posit that the concept of numerosity is fundamentally spatial—a mental number line on which values are envisioned from small to large, much as a ruler shows distance. Once that mental landscape is established, one can determine relative value by comparing locations on the imaginary line. In this manner, one can mentally record and remember that 2 items are a bit fewer than 3 items and that 6 items are a lot more, regardless of the items. Distance along the line becomes a mental analog for abstract number, much as an analog clock depicts time.

In one respect, however, this early, precounting mental number line does not resemble a ruler; this difference has to do with the way people think about quantity when they do not count. For example, without counting, one cannot determine exactly how many birds are in Figure 2.1; at best one can match up the flocks, bird for bird, to see whether any birds are left over—a time-consuming and impractical process. Thus the second key feature of primitive quantification is that it is approximate.

If the uncounted comparisons cannot be precise, how accurate are they likely to be? To answer that question, we return to Figure 2.1: Which decision was easier, A versus B or B versus C? Most people will find the first comparison to be easier. Approximate comparisons are governed by a psychophysical principle called Weber's law. According to Weber's law, accuracy depends not only on the size of the values, but also on their difference: The closer two values are to each other, the harder it is to tell them apart. In other words, the ability to approximately distinguish two values from each other depends upon their ratio: the more similar the values (i.e., the closer their ratio is to 1), the more difficult the distinction. Flocks A and B have 3 and 6 birds, respectively; thus their ratio is 1:2. Flock C has 7 birds, making the ratio of flock B to flock C equal to 6:7—a value much closer to 1. Therefore, distinguish-

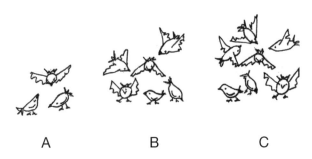

A   B   C

Figure 2.1   Without counting, can you tell whether Flock A is the same size as Flock B? What About Flock B and Flock C?

Figure 2.2.   Without measuring, can you tell which line is longer, X or Y? What about Y versus Z?

ing 6 birds from 7 birds is more difficult than distinguishing 3 birds from 6 birds, when counting is not an option. The larger the collections, the greater the disparity between them must be for someone to detect a difference without counting. If a 6-bird flock lost 3 members, one would notice the difference; if a 100-bird flock lost 3 members, it would be undetectable. (By contrast, errors in exact counting arise when one loses track; the higher the numerosity, the greater the chance of making an error. Thus, exact counting error is simply and directly related to numerosity, not to a ratio of numerosities.)[5]

Weber's law applies to any sequential ordering along a discrete or continuous dimension, such as the alphabet or linear distance. For example, without counting, consider which letter is closer to the letter *O* in the alphabet: *L* or *B*? What about *M* versus *R*? Regarding Figure 2.2, which line is longer? Is it X or Y? How about Y versus Z? Weber's law affects how people compare all kinds of ordered things: the months of the year, weight, color, musical pitch, and numerosity.[6]

As children amass, sort, and distribute collections of things during the toddler, preschool, and kindergarten years, they realize that the sizes of their collections vary and develop a subjective impression of ordered numerosity. Unlike the usual, evenly spaced (i.e., linear) number line that looks like a ruler, this earlier mental image reflects children's greater familiarity with the small values that they use frequently, know well, and can envision sharply and distinctly. By contrast, seldom-used larger quantities are much murkier to young children, as it is harder to tell the difference between two large sets of items if one cannot count them. Larger and less familiar quantities seem less distinct and are therefore more difficult to compare. Preschoolers, for example, can distinguish four items from two but not from six items. A set much larger than about five just seems like "a lot" to a preschooler; finer distinctions are not yet possible. Thus, the early subjective mental number line looks peculiar, with low values spread out at one end and the larger ones bunched up indistinguishably at the other end. Mathematically speaking, these values are arranged more or less logarithmically rather than linearly, as shown in Figure 2.3.[7]

Saul Steinberg's classic "View of the World from 9th Avenue" imaginatively illustrates this youthful mental number line. It depicts a myopic New Yorker's perspective looking west: The artist renders the familiar neighborhood along New York City's Ninth and Tenth Avenues in great detail, much the way people clearly

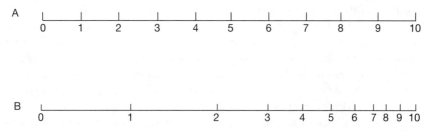

Figure 2.3.   Number lines. A) Linear. B) Logarithmic.

"see" the small, frequently used numbers. Meanwhile, New Jersey, the rest of the continent, the Pacific Ocean, and Asia occupy the leftover space. The view does not offer a sharp geographic distinction between other cities, although it does vaguely represent the general terrain. Similarly, large numbers are out there somewhere in a child's mind, but their exact locations are unclear.

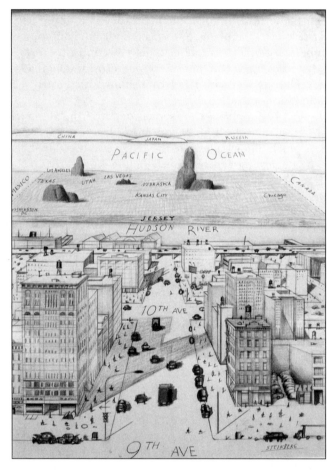

"View of the World from 9th Avenue" © The Saul Steinberg Foundation/ Artists Rights Society (ARS), New York; reprinted by permission.

When exact enumeration is not an option, the only possible view of quantity is approximate and relative. Cultures lacking a counting system can describe and remember collections of objects only approximately, with a Weber-like error pattern mirroring that of very young children.[8] But what happens to these capabilities when people learn to count? Counting—a uniquely human faculty tied to language— allows one to enumerate and remember precisely and absolutely, and opens the door to exact mathematics. For example, by counting, one can describe and remember the exact number of birds in each flock of Figure 2.1; one can also say with certainty which flock is biggest. In some ways, then, there are two entirely different ways to understand quantity. The relationship between these two quantitative systems— the relative and approximate versus the absolute and exact—is one of the most perplexing issues for cognitive scientists.

Evidence of an embryonic mental number line is seen early in children learning how to count. Even 2½-year-olds, when shown a row of three or four objects, will add to or take away from it at one end rather than in the middle, suggesting they think of amounts as a progression.[9] Once children have mastered the basic counting principles, around age 5 or 6 years, they can compare small, familiar quantities (e.g., 2 versus 5), estimate small sums, and enter the number sequence without counting up from 1.[10] This earliest view of numbers is still highly subjective and zero-centric, however. When young children are asked to place a number on a physical number line with an end point that baffles them, they often simply count up from zero, sometimes making hatch marks as they go—just as some New Yorkers may regard the whole world from the viewpoint of Ninth Avenue. Children comprehend small, recognizable numbers, but any forays into the unfamiliar territory of higher values generally produce the errors predicted by Weber's law.

As children gain experience with larger numbers and the counting principles, their mental number line begins to look more linear and they start to manipulate numbers more accurately. In a series of studies, researchers gave pencil-and-paper number lines with only a 0 on one end and 100 on the other end to groups of kindergarten, first-grade, and second-grade students. The children were asked to mark where they thought certain numbers belonged, plotting each number on its own separate number line. The response pattern was logarithmic for the youngest children but became increasingly linear, and hence more accurate, for the older ones.[11] Second-, fourth-, and sixth-grade students produced even more dramatic results with a 1–1,000 number line. Most second-grade students and about half of the fourth-grade students produced a logarithmic response pattern, whereas the older children's estimates were robustly linear, as illustrated in Figure 2.4.[12] The investigators then wondered if the number line range might have influenced the children's accuracy. So they asked second-grade students to mark where certain numbers should go on 0–100 number lines. As predicted, their responses were roughly accurate. The children were then asked to locate those same numbers (i.e., all less than 100) on 0–1,000 number lines. In this larger, less familiar numerical neighborhood,

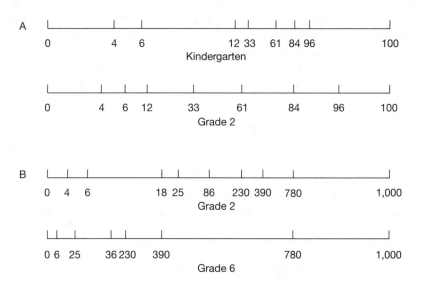

Figure 2.4. Estimated number placements. A) On a 0–100 number line, the results from kindergarten versus second-grade students. B) On a 0–1,000 number line, the results from second- versus sixth-grade students. (*Sources:* Siegler & Booth, 2004; Siegler & Opfer, 2003.)

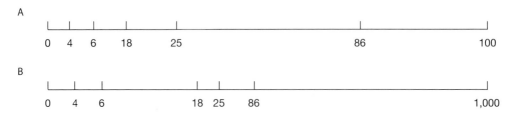

Figure 2.5. Estimated number placements of second-grade students on A) a 0–100 number line and B) a 0–1,000 number line. (*Source:* Siegler & Opfer, 2003.)

the children lost their bearings and produced estimates that assigned outsized proportions to a few relatively small numbers, as shown in Figure 2.5.[13]

In third grade, children begin to think proportionally about the number line.[14] As children learn to parse the numerical landscape and reason proportionally, they can distribute numbers more accurately between the end points of bounded number lines. They also learn to use round numbers to moor their estimations. Going back to the worldview of the myopic New Yorker, just as a few major cities and landmarks protrude from this map, the familiar round numbers (5, 10, 20, 50, 100, 1,000) stand out on the mental number line. They come into focus more clearly than the surrounding countryside and thus can be used to locate other numbers. For example, many people learn history by using key dates (1066, 1492, 1776) to orient themselves on the time line.[15] In the same way, older children and adults often use such landmarks as 250, 500, and 750 to anchor estimates on a 0–1,000 number line.[16] This manner of thinking is crucial to learning fractions, which depends on the idea that the number line can be apportioned. With no natural counting sequence, fractions derive their order and relative values from their places between 0 and 1 on the number line. By fifth or sixth grade, children can begin to track both apportionment and numerical comparison to locate fractions on a number line.[17]

Just as travel can broaden one's perspective, so children's intellectual excursion into the world of numbers can hone their quantitative sensibility, particularly through their exposure to activities related to the number line. Children's measurement estimates (e.g., "If this line is 1 inch long, draw one that is about 5 inches long"), numerosity estimates (e.g., "Guess how many candies are in the jar"), and number categorizations (e.g., "Is this number big or small?"), like their number-line placements, all start out logarithmically distorted and become more linear with age and experience. In studies, children who were good at one kind of estimation task tended to be good at the others, suggesting that estimation skills all rely on a single mental representation of quantity.[18] (Interestingly, preliminary findings suggest that girls may lag somewhat behind boys in developing number-line skills. The reason for this is not known and the results have not yet been replicated.[19])

Although familiarity with the numerical neighborhood is key to understanding mathematics, people typically develop the skill only so far. Unless one is an astronomer or works at the U.S. Office of Management and Budget, even most adults do not truly understand that 1 billion is only 1/1,000 of 1 trillion (see Figure 2.6). *A billion* and *a trillion* are simply synonymous with *a whole lot* to many people.

Figure 2.6. To many people's surprise, 1 billion is only 1/1,000 of 1 trillion.

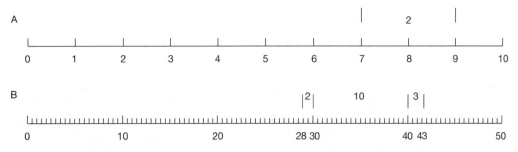

Figure 2.7. Arithmetic on the number line. A) 9 − 7 = 2. B) Decompose 15 to visualize 28 + 15 = 28 + 2 + 10 + 3 = 43, or 43 − 15 = 43 − 3 − 10 − 2 = 28.

Just as people can visualize numerical relations, so can they conjure arithmetic operations visually on a line. One can imagine arithmetic as distances between two points (subtraction) and increments along the line (addition), as illustrated in Figure 2.7, and as repeated increments (multiplication) and equal apportionment (division). Of these, subtraction translates most easily to a spatial analog because it simply involves comparing two points. Although demonstrating multiplication on a physical number line as repeated addition is a valuable teaching tool, the values quickly become too large and complex for the number line to be useful as a mental template on which to routinely conduct the operation. For this reason, children usually learn the multiplication table verbally.

Preliminary efforts to develop kindergarten math screening tests have consistently found that performance on questions of numerical comparison, on and off the number line, was one of the strongest predictors of math achievement in the first few grades.[20] Throughout the elementary years, children's math achievement has been linked to number-line skills and estimation ability in all its applications.[21] It is clear that math ability depends on one's grasp of the most fundamental concepts: numerical values and their relationships.

For most people, ideas about number mature; they learn to estimate reasonably and calculate precisely. However, that does not mean that they never again think of number subjectively. In fact, Weber's law is always lurking; people are most likely to succumb to it when they are in a hurry or cannot count or calculate. Often in daily adult life, people need to solve a problem quickly or estimate the size of a crowd; on those occasions, they tend to draw on their subjective impressions. For example, answer this quickly without calculating: Is it true that $4 + 13 \neq 60$? What about $4 + 13 \neq 19$? Most people find the first question easier to answer than the second because the target number is farther from the true sum—an example of Weber's law at work.[22]

Number and the Brain

How does the brain actually code quantitative information? Advances in imaging techniques that can map brain activation during mental activity provide insight into how the brain "knows" number. Studies of individuals engaged in mathematical thought show that the brain's surface (cortex) becomes active in part of a channel on each side (hemisphere), known as the intraparietal sulcus (IPS; Figure 2.8). The IPS (plural: intraparietal sulci) are highly sensitive to number, regardless of whether it is presented in spoken ("six") or written (*six*) word format or as a numeral (*6*). Moreover, most scholars also agree that the IPS activate in response to concrete nu-

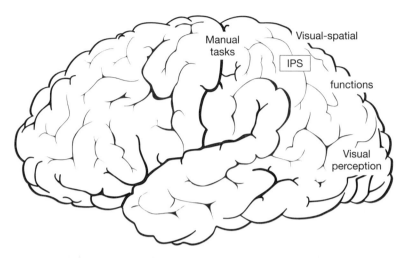

Figure 2.8. View of the brain's left hemisphere, showing the quantity-sensitive intraparietal sulcus (IPS) and surrounding visual-spatial region, as well as the areas activated by incoming visual perceptions and by manual tasks. These functions also reside in the right hemisphere. (*Sources:* Ansari, 2008; Jordan, Wüstenberg, Heinze, Peters, & Jancke, 2002; Simon, Mangin, Cohen, LeBihan, & Dehaene, 2002.)

merosity (******) (e.g., when people compare arrays of dots without counting). As discussed in Chapter 6, the IPS become more sensitive to number as children develop. The IPS encode number only approximately, however, and have difficulty distinguishing two values with a ratio close to 1.[23] The IPS are most active when people compare or estimate quantities, subtract one quantity from another, find the midpoint between two numbers, judge the relative proximity of two values to a third value, and manipulate quantities—that is, when people engage in tasks that can be easily envisioned on a number line.[24]

Significantly, the IPS are embedded in brain regions dedicated to a wide variety of visual-spatial functions. These regions are part of the spatial (i.e., "where") visual pathway connecting visual perception to muscle activities, particularly those involving the hands, and processing information about objects' location.[25] (The "what" visual pathway, also variously referred to as the iconic or object pathway, by contrast, specializes in identifying objects, as well as distinguishing color and shape, and is routed elsewhere.) Difficulty with quantitative manipulations is linked to various medical conditions, such as Gerstmann and Turner syndromes, characterized by damage in this region.[26] Preliminary behavioral data from research in progress suggests a likely association between spatial skill and number-line placement accuracy among typically developing primary-grade students.[27] Moreover, as discussed in Chapter 6, IPS abnormalities have been linked to math learning disabilities in children.[28]

Indeed, several behavioral studies of individuals with brain injuries strongly suggest that people think of numbers and their relationships in visual-spatial terms, as if they could actually see them lined up from small to large along a real line in their minds. For example, one study examined individuals with right-hemisphere brain injuries resulting in hemianopia. In this condition, people lose awareness of their left visual field; when asked to draw figures, they tend to minimize or omit the parts that would be seen to the left of center. The individuals in this study revealed a marked inability to name the midpoint between two numbers (e.g., naming the midpoint between 11 and 19 as 17), even when they did the problem without paper or pencil and responded orally. That is, they committed mental errors similar to the

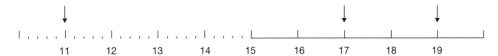

Figure 2.9. A patient with hemianopia responded, "Seventeen," when asked, "What is the midpoint between eleven and nineteen?" (*Source:* Zorzi, Priftis, & Umiltà, 2002.)

more concrete ones such individuals typically make when asked to mark the midpoint of a horizontal line on paper.[29] Figure 2.9 illustrates what the verbal error of an individual with hemianopia might look like on paper and, presumably, in the mind.

It may seem that all spatial sense is necessarily tied to vision, and research often conflates the terms *spatial* and *visual-spatial*. The distinction between them with regard to number sense is not clear, however. A potentially fruitful line of research on the role of purely spatial ability would be to investigate the mental representation of number in individuals who were born blind. One aspect of congenital blindness is that the knowledge of quantity is restricted to sequential auditory and tactile input with the consequent additional memory burden, or to small numerosities of hand-held objects. Young blind children do not use their fingers for counting, but instead use a double-counting system when counting aloud and estimate the results of simple calculations based on a sense of "many-ness" obtained tactilely.[30] More research into vision's relevance to the development of number sense is warranted.

Number Sense and Number Nonsense

Through instruction and experience with quantities, children become familiar with numbers and develop a reliable mental picture of how they relate to each other. A firm grasp of relative quantity fosters more varied problem-solving strategies, more courage to estimate, better judgment about the reasonableness of a solution, and even easier fact retrieval.[31] This confidence with quantities and their mental manipulation can be regarded as *number sense*. As psychologist Ann Dowker remarked,

> To the person without number sense, arithmetic is a bewildering territory in which any deviation from the known path may rapidly lead to being totally lost. The person with number sense . . . has, metaphorically, an effective "cognitive map" of that same territory, which means such deviations can be tolerated, since the person can expect to be able to correct them if they cause problems and is unlikely to become lost in any serious sense.[32]

Most—but not all—children bring an intuitive quantitative sense with them on their first day of kindergarten.[33] For many of those who do start out with a rudimentary grasp of quantity, however, digits often become disembodied from their values and technical arithmetic skills become unmoored from number sense in the course of conventional mathematical education. Sometimes, students rediscover their number sense through computational simplification, like rounding, late in elementary school.[34] However, some students never regain their quantitative intuition and continue through life performing calculations mechanically, without any real idea of what the computations mean. Other students move easily between exact calculations and intuitive approximations, enabling them to tackle unconventional problems, use time-saving short cuts, and devise creative solutions. (In this regard, boys

on average again seem to have an advantage, at least in adolescence; psychologists are trying to find out why.[35])

Like hapless Calvin, some people either lack number sense or do not use it; in fact, most people seem to favor exact—but often poorly understood—numbers.[36] Businesses understand this concept (indeed, they bank on it) when they price $10 items at $9.95, knowing that many customers will be oblivious to the proximity of the latter price to the former. Weak understanding of what numbers actually mean and how they relate to each other—call it *number nonsense*—creates a disadvantage at school and in life when one must verify that calculation results are reasonable or understand a computation's implications.

When children fail to develop number sense, it may lead to other serious math problems. Studies of primary-grade students with severe math impairment found that they were slow to make numerical comparisons (e.g., answering "Which is bigger, five or four?") or made immature number-line placements—both hallmarks of poor number sense. Their slow recital of the counting sequence also suggested a weak grasp of number order, and many had significant difficulty on all other number tasks as well.[37] Tenuous number sense may explain why some young students fail to add using the minimum addend strategy, a common early addition method whereby students determine the larger addend and count up by the smaller term (e.g., $6 + 3 =$ "six . . . seven, eight, nine"). These students may simply not be able to decide which number is the bigger addend or to enter the number sequence at any place other than "one." Because of these difficulties, many such students also fail to master arithmetic facts.[38]

Whereas number sense is closely related to spatial skill, it appears to be quite distinct from some other cognitive skills, such that impairments in one do not imply impairments in the other. For example, some children with receptive and expressive language impairments or dyslexia are able to compare values even if they cannot name or read them, presumably because they understand the relation of one quantity to another.[39] Thus, having a viable mental number line may be regarded as a cognitive function at least partially independent of some other learning-related skills, and compromised number sense may account for some children's math learning difficulties. In fact, severe mathematical disability with no other learning difficulties is not unusual. Most large-scale studies have identified groups of students who have math impairments but no other significant learning disabilities. Future behavioral and neurocognitive research promises to further elucidate the relation between some mathematical disability and impairments in number sense, in the neural circuits that support it, and perhaps in spatial skill.

Difficulty with number lines + + + + + + + + + + + + + +

Emily is a lively 13-year-old with an outgoing personality. Her birth and early development were typical and milestones were on schedule. Now, in seventh grade, her math teacher reports that she seems to lack sense of what numbers mean. She calculates inefficiently and entirely by rote without understanding what she is doing or why.

When asked to place a series of numbers on individual 0–100 number lines, Emily showed the perspective on numbers typically seen in kindergartners and first graders, in which small numbers seem bigger than they are.

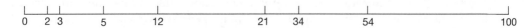

Emily is even less at home in the world of larger numbers, with which seventh graders should be familiar. In locating numbers between 0 and 1,000, she nearly filled the range with numbers less than 100, and she was baffled about the relative values of numbers greater than 100.

Without a reliable, realistic mental representation of ordered quantity, Emily is left with only rote procedures, making mental math especially challenging. For example, when asked (in language she could understand) to mentally subtract 3 serially, beginning at 35, she slowly responded, "Thirty-five, thirty-two, thirty, thirty-six, thirty-three, zero."

Emily received very little prior math instruction using the number line. She learned to calculate using a form of counting, and her teacher's efforts to get Emily to try another method have been unsuccessful. Studies of the effectiveness of intensive number-line instruction to establish a reliable mental number line have focused on much younger children. Testing revealed that Emily's spatial skills are adequate, so it is possible that she would benefit from such instruction.

Classroom Implications

The most difficult aspect of teaching math to young children is to keep their number sense alive and to foster a connection between it and conventional mathematics. Because number sense seems to depend on a reliable mental number line, researchers are now exploring concrete number-line activities, which represent numbers physically as they are represented mentally, as a potentially effective way to teach students about number.

Number lines are not new to the classroom. Teachers have long used number lines created from common objects, including pencil and paper, plastic strips, cardboard tubing, bead strings, linear numerical board games, rulers, calendars, and thermometers—not to mention the chalkboard number line at the front of the classroom. One popular device consists of rectangular sticks of graduated lengths—referred to here generically as *number blocks*—that can be lined up along a number line or track. More recently, computerized versions of these activities have become available.[40]

Like discrete manipulatives, such as buttons or beans, these linear analogs teach cardinality, ordinality, and equality; they can be introduced in tandem with conventional notation and algorithms. Unlike buttons and beans, however, a number line provides fixed visual images (analogs) for potentially unlimited values and is ordered and systematic; moreover, it represents number the way the brain does, linking counting to linear measurement.[41] Number lines can also illustrate the conceptual underpinnings of nearly all elementary number skills, including inequality, numerical comparisons, arithmetic transformations, fractions, decimals, measurement, and negative numbers. Buttons and beans represent quantity as we most often encounter it in life, as random collections; the number line provides a schematic image or mental template that children can rely on and abstract from.

Research with young students suggests that employing number lines in the classroom can significantly improve numerical understanding. For example, helping kindergartners sort the numerals 1–100 into equal-sized piles of *very small, small, medium, big,* and *very big* numbers effectively provided the students with a linear (and thus more accurate) sense of numbers in that range.[42] First-grade students improved their understanding of missing-addend problems using number-line based instruction incorporating practice and feedback.[43] First-grade students also improved in both number-line accuracy and addition skills using computerized number-line illustrations of addends and sums.[44] Instruction using unmarked number lines enabled third-grade students to develop both flexible strategies and procedural competence with multidigit addition and subtraction.[45] Simply correcting second-grade students' most discrepant number placement (typically 150 on the 0–1,000 number line) produced particularly dramatic recalibration.[46] (Researchers do not yet know whether this technique is as effective with children whose view of number is significantly impaired.) Teaching college students to diagram analog-friendly word problems with number lines significantly improved the students' accuracy—more than simply rewording the problems did—and even improved performance on more complex problems.[47] Unfortunately, we have not yet seen any research on the effectiveness of number line pedagogy for children in the intermediate, middle, or secondary grades; such investigations are certainly warranted.

Many preschool and kindergarten children from low-income families without access to certain types of board and other number games have difficulty answering questions such as, "Which is bigger: five or four?"—a quantitative task that poses little trouble for other children of the same age. For these children, number-line activities such as linear numerical board games, which help them link the quantities indicated by spaces counted along a line, dots on dice, numbers on spinners or cards, distance moved, and duration of play have proven particularly useful. Researchers caution, however, that children should count out their moves according to the spaces on the board (e.g., "seventeen, eighteen, nineteen . . . "), not according to the value on the dice or spinner (e.g., "one, two, three . . . "), to connect the counting sequence to the number line.[48]

In one study, four 15-minute sessions using a simple linear numerical board game brought one group of underprivileged preschoolers up to the level of their middle-class peers in terms of number-line estimation, magnitude comparison, counting, and numeral identification—gains that remained 9 weeks later.[49] Another research-based program for children in prekindergarten through second grade, focusing on integrating traditional object counting with diverse child-friendly number-line activities, met with similar success. In addition to numerical board games, these activities included using thermometers and counting off as children queued up for recess—simple daily classroom activities that teachers can implement easily.[50] Explicit number-line instruction led to significantly greater improvements than did nonnumerical board games, counting activities such as card games, prereading intervention, textbook number-line illustrations without instruction or feedback, or traditional math lessons that did not use number-line activities or stress relative magnitude. In short, simple number-line techniques have proven exceptionally fruitful for young children at risk of math difficulties. Alert teachers can find additional linear counting activities in many common settings (e.g., counting steps in flights of stairs or rungs traversed on the monkey bars; playing hopscotch).

The success of early number-line instruction suggests that interventions based on number-line thinking may also prove effective in keeping children's number sense alive while they master arithmetic algorithms. For example, when a child forgets to carry (e.g., $19 + 6 = 15$), a leading conceptual question (e.g., "When you add something to 19, would you expect the answer to be bigger or smaller than 19?") may be more useful than an admonition focused on procedure (e.g., "Don't forget to carry!" or "Check your work!"). We have not seen any rigorous investigation of the effectiveness of such number-sense interventions versus those oriented to arithmetic procedures; such research would provide useful guidance for teachers.

One function for which a number-line approach has proven particularly useful is big-number subtraction, where it helps students avoid some of the pitfalls associated with using the algorithm.[51] For example, when asked to mentally subtract 3 serially from 35, a student struggling to use the algorithm might respond "thirty-five, thirty-two, thirty-nine, thirty-six, thirty-three . . . ." In contrast, the transition across 30 poses less of a stumbling block if the student can envision the subtraction on a number line.

Earlier, it was noted that some students rediscover number sense when they learn rounding rules and computational estimation late in elementary school. Computational estimation must rely heavily on number sense if it is to be useful in daily life. Many people do not estimate very well because it is a complicated task involving approximating a set of numbers, keeping track of all the estimates, doing mental arithmetic with them, and adjusting the results to compensate for the simplifications. Most students take a long time to master this mental juggling act, often well into high school.[52] Nevertheless, one of the key components of this skill—a thorough familiarity with numbers and their number-line neighborhoods—is accessible to much younger children. The typical classroom expectation is for all answers to be precise and for rounding to follow certain rules, but that often causes many cautious students to be less tolerant of approximations as they get older. Conversely, if students develop an intimate knowledge of the number line at a young age, they may be ready for the more complicated estimation tasks later. For example, students can learn to recognize "friendly neighbors"—numbers they can call on, such as 20 or 25, when unwieldy numbers such as 23 are giving them trouble. Further research on the effect of early number-line instruction on downstream computations is certainly warranted.

Although number-line instruction looks very promising, many pedagogical questions remain unanswered. For example, little is known about the relative effectiveness of manipulatives, such as cardboard tubing, versus paper-and-pencil number lines versus computer software engineered to illustrate the same lessons. Each approach has advantages: Objects feature a tactile component, software permits interactive and speeded exercises designed to foster skill fluency, and paper and pencils are cheap and accessible. The relative effectiveness of marked versus unmarked number lines is also unknown, although preliminary research suggests that unmarked number lines may be more engaging and foster development of more variable strategies, even for students with impaired math skills.[53] The effect of color (an iconic feature) should also be tested: When number blocks are color coded, as most are, are students distracted from mastering the necessary association between number and length? Which number-line direction is more effective, vertical or horizontal? Do students have trouble switching from one to the other? Does the mental number line extend to negative numbers? What is the most effective way to teach negative numbers?[54] These questions remain largely unanswered.

Because number-line activities are chiefly nonverbal, some learning disability experts tout their particular usefulness in teaching students with language and reading disabilities; however, that claim has not been scientifically verified. More research is also needed on whether and how number-line instruction might be effective with older students and students whose impairments are spatial or visual rather than linguistic. In general, further research is warranted on the effectiveness of number-line instruction at all grade and ability levels and for students with a variety of cognitive profiles.

Conclusion

Long before most children see the inside of a formal classroom, they know something about number. They know that three cookies are more than two cookies and that if someone takes one cookie away, there will be fewer. They have vague ideas about *a lot* and *a little* that become sharper as they get older. As children learn to count and gain experience with numbers, they develop a mental image of how quantities relate to each other: a mental number line on which each number has its place, like inches on a ruler. Indeed, quantities are coded in the region of the brain that specializes in spatial functions; knowledge of number is intimately tied to that spatial sense.

Many children develop an easy familiarity with quantity—*number sense*—as they gain mathematical experience. With a reliable mental number line, they have a cognitive map that keeps them oriented as they wander through the unfamiliar terrain of school mathematics. Some children, however, lose their early comfort with quantity as they struggle to master arithmetic rules and procedures; others seem never to develop this comfort with numbers at all. It is not yet known why this is so, but research suggests that spatial difficulties may contribute to some children's number problems. For young children at risk for math failure, classroom number-line activities have demonstrated improvement in number sense. Chapter 3 continues the investigation of the relationship between spatial skill and math, looking beyond number to other mathematical branches and applications: geometry, way finding, and proportional and mechanical reasoning.

Notes

[1] Albert Einstein, as cited in Dehaene, 1997, p. 151.
[2] See, e.g., Andersson, 2003; Davis & Pérusse, 1988; Devenport, Patterson, & Devenport, 2005; McComb, Parker, & Pusey, 1994.

[3]See, e.g., Mix, Huttenlocher, & Levine, 2002a; Simon, 1997; Wynn, 1998.

[4]See, e.g., Barth, LaMont, Lipton, & Spelke, 2005; Brannon & Van de Walle, 2001; Mix, Huttenlocher, & Levine, 2002b.

[5]See, e.g., Dehaene, 1997, pp. 64–88.

[6]See Barth, Kanwisher, & Spelke, 2003; Gevers, Reynvoet, & Fias, 2003; Holloway & Ansari, 2008; Moyer & Landauer, 1967.

[7]Huntley-Fenner & Cannon, 2000; Siegler & Booth, 2005.

[8]Frank, Everett, Fedorenko, & Gibson, 2008.

[9]Opfer & Thompson, 2006.

[10]See, e.g., Dowker, 1997; Resnick, 1983.

[11]Siegler & Booth, 2004.

[12]Siegler & Opfer, 2003.

[13]Siegler & Opfer, 2003. See also Ebersbach, Luwel, Frick, Onghena, & Verschaffel, 2008.

[14]Petitto, 1990; Siegler & Opfer, 2003.

[15]See Dehaene, 1997, pp. 108–117.

[16]Siegler & Opfer, 2003.

[17]Rittle-Johnson, Siegler, & Alibali, 2001.

[18]Laski & Siegler, 2007.

[19]Jordan, Kaplan, Locuniak, & Ramineni, 2007; Thompson & Opfer, 2008; J. Opfer, personal communication, February 25, 2008. See also Mills, Ablard, & Stumpf, 1993.

[20]Chard et al., 2005; Jordan et al., 2007; Mazzocco & Thompson, 2005.

[21]Booth & Siegler, 2006.

[22]Ashcraft & Fierman, 1982; Ashcraft & Stazyk, 1981; Booth & Siegler, 2006; Hobbs & Kreinovich, 2006.

[23]See, e.g., Ansari & Dhital, 2006; Cantlon, Brannon, Carter, & Pelphrey, 2006; Kadosh, Kadosh, Kaas, Henik, & Goebel, 2007; Piazza, Pinel, LeBihan, & Dehaene, 2007. But see Shuman & Kanwisher, 2004.

[24]See Dehaene, 2000.

[25]See Culham & Kanwisher, 2001; Jordan, Wüstenberg, Heinze, Peters, & Jancke, 2002; Simon, Mangin, Cohen, LeBihan, & Dehaene, 2002.

[26]See, e.g., Lemer, Dehaene, Spelke, & Cohen, 2003; Molko et al., 2003.

[27]Opfer, Young, & Krasa, 2008.

[28]Price, Holloway, Räsänen, Vesterinen, & Ansari, 2007; Rotzer et al., 2008.

[29]See, e.g., Zorzi, Priftis, & Umiltà, 2002; see also Göbel, Calabria, Farné, & Rossetti, 2006.

[30]Ahlberg & Csocsán, 1999, p. 54.

[31]Canobi, Reeve, & Pattison, 1998.

[32]Dowker, 1992, p. 52.

[33]Jordan, Huttenlocher, & Levine, 1992.

[34]See LeFevre, Greenham, & Waheed, 1993.

[35]Gallagher et al., 2000.

[36]See Hanson & Hogan, 2000; LeFevre et al., 1993.

[37]Geary, Hoard, Nugent, & Byrd-Craven, 2008; Landerl, Bevan, & Butterworth, 2004.

[38]Griffin, Case, & Siegler, 1994; Robinson, Menchetti, & Torgesen, 2002.

[39]Donlan & Gourlay, 1999; Landerl et al., 2004.

[40]See Stern & Stern, 1971; Stevens & Schwartz, 2006; Wilson, Dehaene, Pinel, Revkin, Cohen, & Cohen, 2006.

[41]Stern & Stern, 1971; Griffin, 2004.

[42]Laski & Siegler, 2007.

[43]Fueyo & Bushell, 1998.

[44]Booth & Siegler, 2008.

[45]Klein, Beishuizen, & Treffers, 1998.

[46]Opfer & Siegler, 2007.

[47]Lewis, 1989.

[48]Siegler & Ramani, 2007.

[49]Ramani & Siegler, 2008.
[50]Griffin, 2004.
[51]Blöte, Van der Burg, & Klein, 2001.
[52]Sowder, 1992.
[53]Gravemeijer, 1994.
[54]Fischer, 2003.

Chapter 3

Math and Spatial Skills

Chapter 2 discussed intuitive mathematics by focusing on quantity, a concept that is fundamentally spatial. In a person's mind, values are represented in an ordered fashion, as if along a line, and are compared like distances. Spatial thinking is at the heart of almost all mathematical thought, however, not just number, and is relevant to most mathematical applications.[1] Being able to translate freely between conventional verbal or symbolic formats and spatial images can facilitate the solving of many types of mathematical problems.

This chapter discusses the general role of spatial thinking in mathematics. First, we explore various spatial abilities that undergird mathematics, with special attention to their earliest development. We then examine problem-solving style—the different ways that people draw on their own perceptual and cognitive resources to absorb information and solve problems. Even though *problem-solving style* refers to mental habits rather than actual abilities, it may play a significant role in the way students absorb information in the classroom. The chapter concludes with the relationship of spatial skill to math problem solving and the issue of teaching spatial skills.

Spatial Skills

For most people, the statistical link between spatial and mathematical skills is relatively weak. Conventional mathematics is largely governed by an elegant, self-contained set of rules that can be mastered without reference to spatial concepts. This is especially true of elementary mathematics, including the basic arithmetic and algebra that are the extent of most people's mathematical knowledge. However, the association between spatial and mathematical skills is significant for older, skilled math students who work at a level that demands conceptualization. Indeed, spatial skills (and their variant, mechanical reasoning) predict math performance so well in these adolescents that some psychologists have recommended adding a spatial test to the examinations required for admission to math, science, and engineering programs at the university level.[2] That is not to say, however, that only high school and college teachers need to be concerned about their students' spatial ability: Mathematical ability has been linked to certain spatial skills among even the youngest primary-grade students.[3] Research suggests that spatial ability relates not so much to mathematical performance per se (which can vary with the way mastery is evaluated) but rather to depth of mathematical understanding and versatility of its application.

The relationship between spatial ability and mathematics has sparked scholarly interest, in part because there is a robust sex difference in favor of males (measured internationally and over time) in some spatial and related abilities, despite parity in general intelligence.[4] Not all girls struggle with spatial and mechanical tasks, of course, and many girls perform exceptionally well. But on average, girls are at a disadvantage. The reason for the sex gap, or for the sex gap in certain related math skills, is not clear, but psychologists generally cite a complex web of biological, psychological, and social contributions.[5]

Mental Rotation and Mechanical Reasoning

Spatial ability is a fundamental cognitive skill by which an individual comprehends the location of and relationships among objects in space; it also concerns objects' movement through space and time. One of the most complex spatial tasks, mental rotation, involves imagining objects turning around or in different spatial orientations. Like the other skills discussed in this chapter, mental rotation has its roots in infancy; by age 3 years, children can rotate different-shaped blocks to fit into matching holes.[6] The skill is integral to much of mathematics and is critical for engineering, physics, chemistry, architecture, astronomy, surgery, and other life and behavioral sciences in which mathematics plays a major role. A robust, cross-cultural male advantage is demonstrated for mental rotation throughout development, even among the strongest students.[7] Researchers are attempting to determine whether sex differences in number-line skills in the early grades may be linked to sex differences in mental rotation and related spatial abilities.[8]

Mental rotation ability is typically measured by a test requiring an individual to select the configurations of attached cubes that are rotated versions of a target configuration (Figure 3.1); mental paper-folding tests are used as well. Two-dimensional versions of the task also have been used, particularly with very young children who lack the concentration necessary for the more demanding three-dimensional task.[9] Of all spatial manipulation activities, mental rotation is the most difficult to talk oneself through. Although many students, particularly girls, will attempt a verbal strategy,[10] mental rotation is the spatial task least associated with verbal skill; as such, it can be considered the purest measure of spatial ability.

Mechanics, an extension of spatial skill, is the science of motion; mechanical inference is "any mental process that allows us to derive information about how things move."[11] As such, it also requires understanding the effects of energy on motion. Examples of mechanical problems include predicting the action of simple machines or the effects of gravity and displacement. Unless one can readily draw on past experience or knowledge of a relevant physical law, the most common mode of solution is a thought experiment or mental simulation—that is, a kind of step-by-step mental rehearsal of the events as they might unfold, often revealed in people's gestures as they solve such problems. Try the examples in Figure 3.2. Successful mechanical reasoning depends on how well one eliminates extraneous information and generates mental pictures that are sequentially analyzable and pertinent to the problem at hand. Notably, even individuals who have been schooled in the relevant mechanical principles often perform better by using mental imagery than by applying those

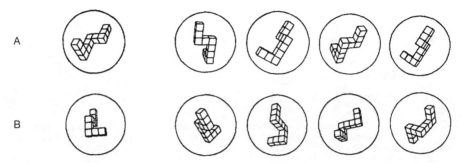

Figure 3.1. Sample items from the Mental Rotations Test. A) The first and fourth alternatives are correct; the distracters are mirror images of the criterion figure. B) The second and third alternatives are correct; the distracters are rotated images of other criterion figures. (From Vandenberg, S.G., & Kuse, A.R. [1978]. Mental rotations, a group test of three-dimensional spatial visualization. *Perceptual and Motor Skills, 47,* 600; reprinted by permission conveyed through Copyright Clearance Center, Inc.)

(often misremembered) rules, suggesting that the two kinds of information are used independently, even though doing so might not necessarily be desirable.[12] In mechanical reasoning, as in mental rotation, males have demonstrated a robust advantage.[13]

Even infants, the most naïve physicists, understand elementary aspects of their world's physical properties and can make some very basic predictions about the motion of inanimate objects. They know, for example, that objects are both cohesive and bounded, even as they move, and that they exist and move in continuous fashion over space and time.[14] Infants also pay attention to an object's source of energy.[15] Are these early ideas about the nature of the physical universe innate? As with numerosity, the location of the developmental headwaters of mechanical reasoning remains the subject of lively scholarly debate. Regardless of their origin, however, these early notions form the foundation for all subsequent learning and reasoning about objects and motion. As infants' sensory and motor faculties mature, they begin to amass observations about objects in motion. Gradually, they form theories

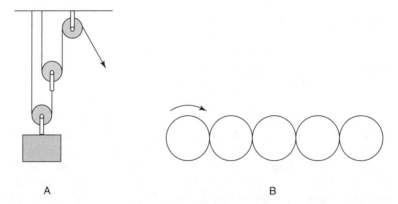

A   B

Figure 3.2. Mechanical reasoning problems. A) When the rope is pulled, will the bottom pulley turn clockwise or counterclockwise? B) When the left gear turns clockwise, in what direction will the far-right gear turn? (From Hegarty, M. [1992]. Mental animation: Inferring motion from static diagrams of mechanical systems. *Journal of Experimental Psychology: Learning, Memory, and Cognition, 18,* 1085 [published by the American Psychological Association]; Reprinted from *Trends in Cognitive Sciences*, Vol. 8. M. Hegarty, Mechanical reasoning by mental simulation, page 281. Copyright 2004, with permission from Elsevier.)

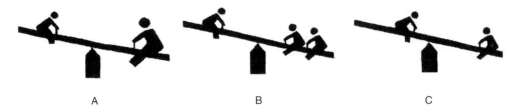

A   B   C

Figure 3.3. Development of balance concepts. Children learn in stages that the side that goes down A) has the heavier weight, B) has more weights, or C) has the weight farther from the fulcrum. They learn much later to consider these factors simultaneously. (*Sources:* Case & Griffin, 1990; Siegler, 1976.)

and then revise them with each unexpected outcome. They also figure out what variables are salient and develop categories of events. For example, infants learn from experience that inanimate objects move not only when something bumps into them, but also when they lose their support and fall. In this way, children become increasingly accurate in their predictions.[16]

Complex mechanical reasoning often requires one to keep several factors in mind simultaneously—a mental juggling act that relies on practice and working memory (a slowly developing mental function; see Chapter 8). Take simple levers, for example—a lesson children often learn on the playground's teeter-totter. Extrapolating from balance beam studies, it is known that at age 4 years, children understand that when they sit on one end and their father sits on the other, Dad (who is heavier) will go down, as shown in Figure 3.3. By age 6 years, children figure out that if two classmates sit on the other side instead of Dad, those friends also make that side go down—that is, they can now make an association between numerosity and weight. Around age 8 years, children begin to take into account relative distances from the fulcrum, although the ability to weigh all three factors together often does not occur until adolescence.[17] Children thus enter the primary grades with some fairly sophisticated, albeit still only implicit, spatial and mechanical skills. These capabilities form the bedrock upon which much of their formal education in a wide variety of fields will rest.

Early Geometry

Early geometry skills, which are closely related to mental rotation, also have their developmental origins in the mists of early childhood. Although they are not conscious, classroom-type skills, an implicit capacity for geometric thought can be inferred from young children's behavior. One such capability is locating objects in space, one of children's earliest geometric skills. By 5 months, infants can use geometric information, such as the angle and distance of an object in relation to themselves and nearby referents, to register and remember the location of an object (e.g., a toy they watched someone hide in a sandbox). By the time they learn to walk, children can factor the distance and direction of their own movement into their object-search strategy—a process called dead reckoning. True geometric skills (i.e., using distance and direction to calculate the spatial relationships among objects) emerge fairly abruptly around 21 months. Because certain sites deep in the brain mature at this age, children gain the capacity to learn the locations of more than one object at a time; the learning itself, however, comes from experience.[18] Sex differences in object search become apparent later. In one study of object search among adults, women remembered the objects' locations as well as men, but men identified the positions, apart from the objects that had occupied them, better than women.[19]

Another development in the second year of life is toddlers' ability to orient themselves in relation to the boundaries of their surroundings—a task they also accomplish using geometric cues. Psychologists discovered that young children who were experimentally disoriented in a featureless room could reorient themselves using the relative dimensions of the surrounding walls and their directional sense.[20] Children learn to use landmarks (e.g., color, other objects) somewhat later. Landmarks increase the accuracy of reorientation for older children and adults; experimental evidence, however, suggests that being able to use both boundary and landmark cues at the same time requires language.[21] Among adults, the ability to reorient oneself in a featureless, enclosed environment is strongly linked to mental rotation skill and shows the expected male advantage.[22]

Geometry demands that we understand not only real-time spatial events, but also diagrams and maps of where things are in relation to each other. The ability to use such schematic analogs emerges surprisingly early in childhood, around age 2 years, when children can find an object in a room after seeing the room's layout from above. By first grade, children understand the proportional concept of scaling, can locate real objects from abstract map symbols, know how to rotate a map to align it with their own perspective, and can use a map to plan routes through a complex environment.[23] Although map reading improves with formal education, as do other geometric skills, even unschooled members of primitive cultures can read and draw basic maps, suggesting that the underlying cognitive abilities are either innate or can be acquired without formal schooling.[24] Route learning, like orienting, has been linked to mental rotation; as such, it has shown a male advantage that is evident by age 5 years. A closer look at route-learning strategy reveals that males do best using the geometric properties of the route (distance and direction), whereas females do better with landmarks. Moreover, females tend to use a less efficient verbal, rather than spatial, strategy to absorb and retain directions.[25]

Understanding and manipulating shapes, the usual subject of geometry lessons, involves both spatial skills (e.g., locating the shapes and their parts in relation to each other) and iconic skills (e.g., describing, comparing, and transforming them). The ability to distinguish and categorize geometric shapes begins early in infancy: Even newborns can distinguish perceptually between different kinds of shapes and are aware of symmetry across the vertical axis.[26] With age, experience, and improved working memory, young children fine tune their shape discrimination, although scientists have found unschooled awareness of geometric properties in cultures with no formal geometric language or training, suggesting the existence of a fundamental "geometry sense."[27] By kindergarten or first grade, most children have a mental working model of circles and squares; they can name them and pick them out of a lineup. Recognizing shapes with more variants, such as triangles and rectangles, may take longer to develop. At first, young children discriminate shapes chiefly by the length and position of their sides; angles are less salient to them and more difficult to understand. Preschoolers use angles in their play, however, and can match them. They also use parallelism and perpendicularity; begin to grasp similarity, congruence, and slide transformations; and use geometric shapes to make pictures. All these developments, however, require instruction to mature and become explicit.[28]

One vital skill for classroom geometry is the ability to copy shapes with pencil and paper. Successful pencil-and-paper reproductions of geometric forms rely on spatial and iconic visual skills to comprehend the relative location and shape of the figures and on fine motor abilities to draw. Graphic replication also depends on the integration of these skills, a brain function that develops with practice throughout childhood. Being able to copy figures correctly has been linked to math achievement.[29]

Difficulty copying designs + + + + + + + + + + + + + + +

Elizabeth is a gregarious 13-year-old girl who has trouble setting her work out on the page. For example, in executing multiplication using an algorithm requiring her to array the operands in a rectangle, Elizabeth's pencil-and-paper production is disorganized — almost as if she does not know what a rectangle is, even though she has seen many examples.

One clue to Elizabeth's paper-and-pencil difficulties lies in her difficulty with copying what she sees. In the following example, she struggled to copy a diamond that would give no trouble to most children her age.

In the next example, Elizabeth copied most of a triangle correctly until she got to the left-most angle. She seemed not to know how to deal with it but finally decided to sneak up on it by small increments along each adjacent side until the lines came together.

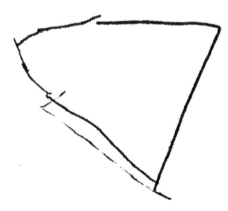

Elizabeth struggled to fix the angles on several copying tasks with much erasing and progressively improved redrawing, suggesting that she perceives the originals accurately. Also, her legible handwriting and good pencil control on other drawings indicate that her pencil skills are adequate. Thus, Elizabeth's difficulty is in *integrating* her visual and grapho-motor skills, not in the skills themselves. Elizabeth's visual-motor integration difficulties are not limited to pencil and paper. Despite adequate mental rotation, she also did not perform well on a test requiring her to copy a two-dimensional design with colored cubes by physically turning and manipulating the cubes until the design on top matched the given one on paper.

Because of these problems, Elizabeth likely will have significant difficulty learning geometry, although she may learn some principles if she can do so using only her eyes and pointing. To our knowledge, there is no research on rehabilitation of pencil-and-paper visual-motor integration difficulties. Future research may examine the effectiveness of using sensory-motor techniques (e.g., copying a design by drawing through an identical stencil or around a tile of the same shape) on design copying specifically and on geometry performance in general.

Fraction Sense

As noted in Chapter 2, infants can distinguish between different amounts of "stuff." They can also distinguish between *proportions* of "stuff." For example, 5-month-olds will notice that there is a difference between identical containers of liquid that are one-quarter versus three-quarters full. Children's "fraction sense," like the other mathematical intuitions discussed in this chapter, begins very early. By age 3 years, typically developing children can pick the bigger of two fractional portions of circles; by age 4 years, they can draw an analogy between equivalent proportions of different geometric shapes. They can also add fractional portions of shapes from memory and use scaling in map reading. As their working memory improves and they gain experience, children's fractional sense broadens. For example, by age 6 or 7 years, children can compare and add combinations of whole and fractional objects and make probability decisions based on proportional amounts. Around age 7 years, children begin to understand that sharing something among four friends gives each person less than would sharing it among two, the concept later necessary to understand the comparison of fractions with different denominators.[30]

Children's fraction sense begins with continuous dimensions, perhaps because it relies on basic perception. (Grasping proportions of discrete sets, such as bags of candies, is not as salient because children have other ways to apportion such things.[31] It is also complicated by counting and by the abstract idea that some objects are subsets of others, as discussed in later chapters.) Although young children do not have the technical terms to explain their thinking, most of them easily recognize proportional relationships in concrete and schematic forms. In this manner, the conceptual groundwork for fractions is laid, quite naturally, long before teachers introduce the notoriously vexing notational conventions in the classroom.

Not all children come by fraction sense easily, however. One study identified a small percentage of middle schoolers who were unable to rank order fractions even when they were depicted simply as fractional parts of circles.[32] Future research is required to determine whether such an impairment in fraction sense, like that in number sense, is attributable to an underlying cognitive impairment, perhaps in spatial ability.

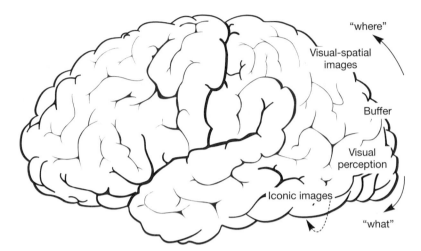

Figure 3.4. Visual perceptions from the eyes enter at the back of the brain, shown here in the left hemisphere. From there, the activation progresses through the buffer to the visual-spatial and iconic imaging areas. (*Sources:* Jordan, Wüstenberg, Heinze, Peters, & Jancke, 2002; Kosslyn, 1994; Ungerleider, 1995.)

The Neurology of Spatial, Mechanical, and Geometric Reasoning

Behavioral observations tell us that even young children understand a great deal about objects, including their shapes, spatial relationships, and movement. But how does the brain code this information? As illustrated in Figure 3.4, incoming visual perceptions, received at the back of the brain, travel to a buffer area that begins to sort and organize them. From there, spatial information is routed to the visual-spatial pathways at the top of the brain, which track location and motion; iconic information is sent along circuits in the lower portion of each hemisphere that identify objects and their properties, such as color and shape.

When forming mental images, a person draws perceptions from memory rather than sensory input, but otherwise, the same pathways are involved for both.[33] Because people tend to prefer one or the other type of imagery, some scholars have speculated that the visual buffer has a limited capacity, making it difficult to integrate iconic and spatial information, such as photographs and maps.[34] Visual tasks activate additional brain areas, of course, including some areas deep inside the brain that code aspects of route-learning.[35]

The spatial pathways activate during such math-related tasks as mental rotation and dimensional perception; spatial memory; judgment of the relative sizes of lines, angles, and geometric shapes; bisection of lines; enumeration of objects scattered across the visual field; distance assessment; route learning; and mechanical reasoning. As discussed in Chapter 2, number sense also resides in this part of the brain; in fact, several brain studies show activation of the intraparietal sulci during mental rotation, lending support to the idea that numerical judgments are essentially spatial.[36] The iconic pathways govern shape recognition and analysis; they also play a major role in reading letters and numerals.[37] Many math tasks draw on both spatial and iconic images; to the extent that someone uses spatial or iconic features in a solution strategy, the brain activates accordingly.[38] Research indicates that even when males and females perform equally well on a mental rotation task, they activate somewhat different cortical areas—thus suggesting they use different strategies.[39]

It is commonly assumed that spatial skills are right-brain functions, but the scientific findings are more complex. Broadly speaking, the right hemisphere processes information globally, whereas the left hemisphere processes information sequentially and analytically. In most people, the left hemisphere controls language, which is, of course, sequential. Hence, any task requiring linguistic input automatically involves the left hemisphere in addition to whatever other functions may be in play. The important factor in determining hemisphere involvement is not so much the nature of the problem as how one solves it. For example, although an individual may be able to mentally rotate a simple configuration in one mental twist, thereby relying chiefly on the right hemisphere, more complex configurations might require sequential analysis or verbal insights, drawing on the left hemisphere.

One study found that individuals with "split-brain" conditions (in which the nerve fibers connecting the two cerebral hemispheres have been therapeutically or traumatically severed) could match visible shapes to hidden forms explored with their hands better with the right hemisphere. These shapes included the regular Euclidean forms (i.e., triangles and squares), as well as less easily analyzable or verbally describable amorphous ones. Looking only at performance on the relatively familiar Euclidean forms, however, the researchers found that the left (language) hemisphere performed nearly as well as the right. As shapes became more irregular and progressively lost their defining characteristics, however, the language hemisphere became increasingly useless. In general, the more one uses sequential analysis and verbal description for a spatial task, the more involved language centers of the brain will be.[40] On some tasks such as mental rotation, however, those who are able to rely solely on the right hemisphere tend to perform better than those who call on both.[41]

Before considering what this all means for mathematics, we examine the research on problem-solving style—the habits of mind that silently inform the way one approaches problems.

Problem-Solving Style

Spatial thinking does not come naturally to everyone. In fact, even for the most straightforward spatial problems, different people use various cognitive strategies. For example, when driving, some people prefer verbal directions, whereas others reference landmarks or imagine the route on a map. Many people use more than one strategy, but this example illustrates an important psychological reality: People absorb and remember information in different ways. Although people usually use several different methods and can choose how they approach problems, many have an unconscious default option—a problem-solving style. These methods are constantly at work in the office, at home, and in school, where they often operate unnoticed by students and teachers alike.

At one time, psychologists thought of people as being simply either verbalizers (i.e., those who solve problems with words) or visualizers (i.e., those who solve problems with mental pictures). However, scholars have since looked more closely at visualizers and found two distinct kinds: spatial and iconic. People who visualize in spatial images think primarily in terms of objects' locations and movements in space. These mental pictures are generally schematic in nature, like a road map or assembly instructions, highlighting only the relevant spatial information such as location, distance, and direction. Because the mental pictures are spare, they are amenable to analysis—that is, they can be mentally taken apart, rearranged, analyzed piecemeal, and even set in motion. The mental pictures of people who visualize in iconic images, by contrast, highlight objects' physical appearance (e.g., shape, color) rather than their spatial characteristics. These iconic images, like photographs or paintings, are detailed, realistic, and typically seen as a coherent whole.[42] These two modes of visualization are consistent with the idea that our brains process visual information via two separate pathways.

Verbalization and visualization are somewhat independent. Those who prefer verbal-logical problem-solving strategies demonstrate a normal range of spatial skill, and visualizers show a normal range of verbal skill. Verbalizers and spatial visualizers both tend to be better at mental rotation than iconic visualizers, but spatial visualizers are faster at it than those who use verbal strategies.[43] Not surprisingly, people perform best when they choose problem-solving methods that match their cognitive strengths. For example, researchers examined how people approach logic puzzles (e.g., "If Adam is taller than Benjamin, and Benjamin is taller than Cain, who is tallest?"). Most of the participating college students used a mix of verbal and schematic strategies. Of those using a purely verbal strategy, however, performance was robustly linked to their verbal ability, whereas the performance of those using a purely spatial strategy related strongly to their spatial ability.[44] For some people, however, a persistent preference for one strategy may mask an impairment in the skill required for the other. Students who show marked reluctance to consider an alternative strategy (e.g., diagramming a word problem) may be doing so because they are cognitively ill equipped to think that way; in those cases, a cognitive assessment may clarify the situation and reduce classroom stress.[45]

Notably, the two visualization modes are quite distinct and are not generally shared by a given individual. Those who excel at one form of visualization often do poorly at the other, and the degree to which people prefer to solve problems visually is linked to their spatial or iconic strengths. Generally speaking, scientists, engineers, and mathematicians tend to rely on spatial images, whereas those who work in the creative arts are more likely to use iconic ones.[46] However, this pattern does not indicate whether preferences are innate or learned; either explanation is possible.

Mathematical Problem Solving

The importance of spatial imagery for engineering, architecture, physics, and other mathematical applications is obvious. The ability to conjure and manipulate spatial images or analyze diagrams (and the habit of mind to do so) is just as important in the ordinary math classroom, however—not only for understanding numerical relations, but also for performing many types of mathematical reasoning tasks. These tasks include distilling and organizing information from word problems, making and following logical arguments, and reading graphs.[47] Likewise, they encompass identifying geometric relations, deciphering complex functions, and interpreting and generalizing solutions. Some psychologists think that the advantage of spatial or schematic imagery over iconic imagery is its analyzability; others suggest that iconic imagery can actually weaken analytical reasoning by cluttering it with irrelevant detail.[48] As the eminent cognitive psychologist Diane McGuinness noted,

> The ability to conjure up mental representations of three-dimensional objects and to imagine them in motion is more than just a bonus for the mathematical mind. In certain cases, it is the essence of the process, the sine qua non without which the symbols in the formula would have no meaning.[49]

Mathematics and Spatial Thought

Whatever the explanation, the importance of schematic imagery and spatial ability to mathematics is evident. In one study in Ireland, researchers investigated the link between visualization preference and mathematical performance in a class of sixth-grade boys. The boys were given a word problem, such as the following:

> A hitchhiker set out on a journey of 60 miles. He walked the first 5 miles and then got a lift from a lorry [truck] driver. When the driver dropped him [off], he still had half of his journey to travel. How far had he traveled in the lorry?

The boys were then interviewed about how they solved the problem. Of the boys who visualized the problem, those who said they envisioned the scenario schematically (e.g., "I didn't see the lorry driver. . . . I just pictured 60 miles. It could have been 60 feet, 60 anything.") did well on the problems, whereas those who reported iconic images (e.g., "I just imagined [the hitchhiker] outside his house with his hand out, hitchhiking") tended to do poorly.[50]

Two ways to envision a problem $+++++++++++$

Cassie is an outgoing 14-year-old eighth grader who seems at ease with adults. Although left handed, she has a strong right handshake. Cassie was brought to our attention because she has difficulty conceptualizing applied math problems and occasionally seems spatially disoriented.

Cassie's birth and early development were unremarkable. She was able to use both hands equally well until she began showing a preference for writing with her left hand in second grade. Although reading and spelling have come easily to her, she has some difficulty with both reading and listening comprehension, particularly in regard to complex stories or instructions. In contrast to her otherwise average-range verbal skills, Cassie's nonverbal abilities are substantially below average (below the 10th percentile for her age). She particularly has difficulty with spatial skills: She scored in the 12th percentile for her age on a test of mental rotation and even lower on a test requiring her to construct an abstract design to match a given one. In presenting a class project on a tri-board display panel, she arrayed her material randomly on the boards with puzzling empty areas and seemed not to realize that something was missing, despite explicit instruction on how to fill up the space.

In math, Cassie is unable to use diagrams to help her solve simple, practical problems. For example, when faced with the problem, "A person sets out to run 7 miles. He runs 2½ miles. How much farther does he have to run to get to 7 miles?" Cassie suggested multiplying 2.5 × 7. Her teacher tried unsuccessfully to explain the problem with a diagram, as illustrated here. Only explicit instruction using a simplified example (running 1 mile of a 2-mile run) helped her to see that subtraction was the appropriate operation.

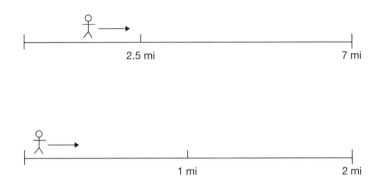

Furthermore, when Cassie does produce an illustration to help her solve a problem, she tends to include iconic details that are irrelevant to the mathematical structure of the problem. She is clearly more comfortable with iconic than spatial (schematic) images, performing in the 95th percentile for her age on a test requiring her to find the missing detail in a drawing of a familiar scene.

The following drawings illustrate her solutions to two word problems. For each she was asked to draw something that would help her solve it. Cassie's drawing is shown on the left after each problem. On the right, for contrast, is the drawing and solution produced to the same problem by her classmate, Craig. Craig's favorite pastimes include video games and playing with LEGOs; in school, he loves hands-on science activities. His scores on tests of mental rotation and design construction measured in the 95th and 75th percentiles for his age, respectively, in sharp contrast to Cassie's on the same tests.

A man wants to visit his mother in another town. To get to his mother's house, he leaves his house and drives straight down the highway for 10 miles. Then he turns right and drives another 4 miles until he gets to a stop sign, where he turns left. His mother's house is 2.5 miles down the road from the stop sign. Draw a map of his trip from his house to his mother's house.

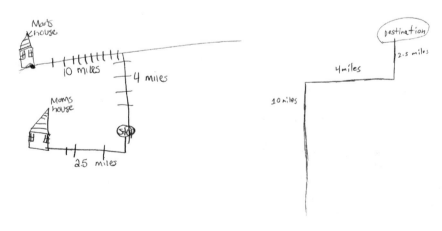

A girl wants to run around a rectangular park. She starts at a corner of the park and runs one mile along one side of the park to the next corner. She turns the corner and runs two miles to the next corner. How many miles will she run altogether if she ends her run where she began it?

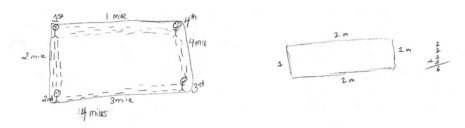

Note that Cassie confuses left and right, loses track of the connection between the math and the diagram, and uses numerous iconic images in her drawings. By contrast, Craig has pared each problem to the bone and has produced spare, accurate schematics of each problem. It is difficult to know whether Cassie's ambidexterity relates to her spatial difficulties. Most generalizations about the functional organization of the cerebral hemispheres come from right-handed individuals; the hemispheric organization of left-handed and ambidextrous individuals is less predictable and may affect their patterns of cognitive strengths and weaknesses.

Diagrams can be especially helpful in clarifying and setting up complex novel problems, especially when one is first learning a concept or when formal symbolic statements are cumbersome or counterintuitive. Sometimes a simple diagram can elegantly illustrate even the most gnarled conventional constructions. While creating a useful image requires mathematical understanding, schematic visualization can also lead to further insight when successful.[51]

Research suggests that individual differences in spatial ability may account for some differences in math ability. One longitudinal study looked at the relationship of scores on a block-building test (requiring both spatial sense and an implicit knowledge of gravity's effects on balance) taken at age 4 years to math grades, math course selection, and standard math test scores in third, fifth, and seventh grades, as well as in high school. The researchers found that although there was no relationship to math achievement in elementary school (where, presumably, algorithmic approaches were emphasized), by seventh grade there was a statistically significant link between the students' preschool block-building skills and their scores on a standardized math test. By high school, the students' preschool block-building skills correlated significantly with their math scores, grades, and course selection.[52] If these results can be replicated in larger groups of students, psychologists may want to consider including a construction-skills assessment for kindergartners who are at risk of math difficulties.

Studies of middle school students who are gifted in math lend support to the link between spatial ability and mathematics. In a 20-year follow-up study of 13-year-old students with precocious math talent, researchers found that superior spatial ability relative to verbal ability (above and beyond math skill) was a strong predictor of career choices in mathematics, architecture, engineering, and computer and physical sciences.[53] Although middle school students with strong math skills do not differ from other gifted students in algorithmic competence, they outperform both verbally gifted and average students in mental rotation. For those students who are not gifted in math, mental rotation is significantly associated with math achievement. Math-gifted students excel at solving novel math problems requiring the ability to envision and mentally transform schematic images. In fact, the ability to translate a verbally presented spatial problem into a schematic drawing and then to put the answer back into words robustly distinguishes those gifted in math from other students, including those gifted in art—a finding supported by cross-cultural studies.[54] One educator has speculated that "gifted individuals have their own internal 'blackboards' and can visualize complicated structures without being aware that they are doing so."[55] Thus, being able to recruit two independent and complementary problem-solving strategies (visual-spatial and verbal-logical) defines those who are gifted in math, at least by fifth grade and beyond.

Not surprisingly, students tend to approach problems differently, depending on their particular strengths and weaknesses. Researchers investigated two groups of middle school students who had significant discrepancies between their visual-spatial ability and their verbal skill. One group had strong visual-spatial skills and weak verbal ability, whereas the other group had the reverse cognitive profile—patterns common to individuals with learning disabilities. Participants were given word problems and were then asked to make a drawing that could help them solve the problems. Although there was no difference between the two groups in the overall number of correct responses on the test, a significant difference existed in the ways the groups went about solving the problems. On interview, students with weak spatial ability and strong verbal ability provided more detailed verbal descriptions of the relevant information in the problem, whereas students with the reverse cognitive profile more thoroughly translated the problem into pictures.[56]

In the long run, does the range of strategies available to a student make a difference? In one study of high school juniors and seniors, those students who relied primarily on algorithms (i.e., verbal-logical strategies) had a more negative attitude toward math than did those who combined algorithmic and insightful strategies.[57] One may speculate that as problems become more complex, the inability to conjure mental images to gain insight would divorce the algorithms from the problems they model and the solutions from their relevance, thereby potentially quashing interest and motivation for some students.

Math, Space, and Sex

The role that spatial ability plays in mathematics learning is particularly salient for some girls, who struggle with it more than do many of their male classmates. The male advantage in mental rotation skill statistically accounts for most—if not all—of the sex differences in the SAT math section scores of the top-achieving students, among whom boys are typically overrepresented.[58] Studies have shown that on items for which spatial approaches are required, both boys and girls use them successfully; on problems for which spatial strategies are optional (but, if used, would provide an advantage in speed or accuracy), boys are more likely than girls to use them and excel.[59] On one international achievement test, the math items that showed the largest male advantage tended to be those amenable to a spatial solution, as in the following example[60]:

> Four children measured the width of a room by counting how many
> paces it took them to cross it. The chart shows their measurements.

Who had the longest pace?	Name	Number of paces
A) Stephen	Stephen	10
B) Erlane	Erlane	8
C) Ana	Ana	9
D) Carlos	Carlos	7

When researchers tested a group of American tenth-grade students with these items, as well as with measures of spatial and mechanical reasoning and math self-confidence, they found that the composite spatial-mechanical ability score accounted for most of the sex difference in the math scores.[61]

To the extent that students execute math problems by conventional logical or verbal strategies, math skill demonstrates no sex difference. In fact, the availability of several nonspatial strategies may help explain the sex parity in achievement among students whose math is limited to basic arithmetic.[62] To the extent that one can solve math problems more quickly and efficiently using schematic visualization strategies, however, girls on average are relatively disadvantaged.[63] As some of the previously cited work suggests, girls may be more in the habit of approaching problems verbally and with iconic images, although more research is needed to verify this. Specifically, future studies should investigate whether girls' performance would improve if they were simply instructed to use spatial strategies whenever possible.

Broadly speaking, girls have more trouble than boys figuring out the mathematical structure of problems. In one high school study, despite comparable achievement, boys were better than girls at identifying the necessary information missing from algebra word problems and were quicker at solving problems containing irrelevant information. The girls more frequently perceived irrelevant information as being necessary for solving the problem and incorporated it into their solution attempts.[64] Other tasks requiring an understanding of relationships, such as physical proportions and verbal and nonverbal analogies, have also demonstrated a male advantage.[65] Superior problem representation may also help explain boys' advantage on SAT math problems with unusual twists. As McGuinness noted, math problems may have more meaning for boys because they experience them "as a representation of something real" that they can more easily envision.[66]

Although boys are better at mental rotation, that skill is more closely linked to math performance for girls than for boys. In the study cited previously that examined the word-problem performance of middle school students with asymmetric cognitive profiles, girls with strong verbal and weak spatial skills needed the most help in drawing the diagrams, whereas girls with the opposite profile needed the least amount of help with drawing. The boys fell somewhere in between.[67] Spatial ability simply makes a bigger difference to girls than to their male classmates; therefore, teachers will want to be alert for girls with weak spatial skills.

Spatial skills affect the academic paths of the most capable girls beginning early in adolescence. In a longitudinal study of academically talented 13-year-old students, researchers found a close connection between their spatial abilities and later subject preferences in high school. Mental rotation was particularly linked to dislikes for girls: The least favorite subjects for girls with good spatial skills were in the humanities and social sciences, whereas the least favorite classes of girls with poor spatial skills were in math and science. High school course experience and SAT score performance often inform first-year college course selection. In math and science, loss of continuity can discourage students from returning to those subjects later. Among the boys in the study, subject preferences were only weakly related to their spatial abilities.[68]

The Relationship of Math Difficulties to Spatial Ability

What is known about the spatial abilities of students with the weakest math skills? A number of studies, collectively covering early childhood through adulthood, have explored nonverbal skills in pupils with math disabilities; all have found that spatial impairments accompany math difficulties in some students. However, most of the studies approached the question of compromised nonverbal functioning in a broad, exploratory manner and did not focus on mental rotation, mechanical reasoning, or other specifically spatial skills per se. Even so, some students with math difficulties scored demonstrably lower on cognitive tests that required at least some two-dimensional rotation.[69]

How math difficulties may be linked to spatial impairments is an urgent topic requiring more research. Spatial ability relates not only to math tasks commonly thought of as spatial (e.g., learning geometry or physics, aligning digits in complex calculations) but to the very concept of number itself and to problem solving in general. A recent study of third-grade students with exceptionally poor addition and subtraction fact mastery found that their only significant cognitive weakness was in their spatial abilities. Many such students continue to rely on fingers and other objects to track values. One must suppose they lack that second port of entry to number—a reliable mental number line.[70] For many students who progress beyond elementary arithmetic to topics requiring deeper conceptual understanding, weakness in spatial thinking and schematic diagramming can interfere significantly with representing and solving complex problems.

Can Spatial Skills Be Learned?

By now, readers may have arrived at two key questions: Is it possible to teach such a fundamental cognitive function as spatial skill? And if so, would it make a difference in the math classroom?

Recall from Chapter 2 the provocative evidence that classroom spatial activities, such as linear numerical board games and direct number-line instruction, can enhance fundamental number sense in young children. Like number-line scholarship, the work on spatial skills training is quite new. However, one research team gathered all the existing evidence (more than 100 studies, most of which have not been published) and concluded that it is possible to train spatial skills. Using training methods as diverse as making dress patterns, playing Tetris, and locating a tilted pitcher's water line, researchers have determined that children and adults can improve their spatial skills. In particular, these techniques are effective with girls and women, allowing them access to educational opportunities and professions that require such abilities.[71]

A few of these studies have produced improvement in meaningful classroom capabilities. A formal 3-month training program focused on the computer graphics skills required of first-year undergraduate engineering students enhanced retention of all students, including women, both in their engineering major and in their graphics-oriented program. Similar specially designed programs boosted all students' abilities and eliminated the male–female achievement gap in a geology course, as well as improved calculus students' performance on solids-of-rotation problems.[72] Although most training methods have not produced parity, researchers have found that the spatial skills of both sexes have plenty of room to grow; thus, both boys and girls can benefit from training. For the mathematics curriculum, more research is needed on ways to encourage and improve spatial strategy and diagram use, particularly among girls, beginning in the earliest grades.[73]

One interesting approach that has offered promise is music instruction. After finding that keyboard instruction improved average-ability, middle-income kindergartners' skills on jigsaw puzzles and block building, researchers examined the effect on those same skills in younger children at risk for math failure. They found that 3- and 4-year-old Head Start attendees with below-average spatial ability who received individual weekly piano instruction during 2 years of preschool performed significantly better than children without piano lessons—not only on jigsaw puzzles, where they scored above average, but on general math skills as well. As late as second grade, their math performance remained at the level of middle-income classmates who had had no musical training. To determine which aspect of music training most enhanced math and spatial skills, researchers provided lessons in either piano, singing, or rhythm to another group of Head Start preschoolers. When tested again in kindergarten, the children who had studied piano and singing performed at the level of their middle-income classmates on spatial-temporal tasks, as well as in arithmetic, mathematical reasoning, and numeracy. Those who had received rhythm instruction, however, surpassed them all, suggesting that it has the strongest impact on cognition and math achievement.[74]

Other researchers found that adding piano instruction to the second-grade curriculum, along with computerized training in visual-spatial skill, significantly improved students' understanding of the relationships of parts to wholes. These investigators attributed their success to students' new appreciation of pitch intervals and to the proportional relationships of rhythmic patterns within fixed measures. They noted, however, that although these young students demonstrated insight into fractional and proportional relationships on concrete tasks, they could not be sure that this understanding would necessarily translate to conventional fraction notation later.[75] Replication of these music studies with larger groups of students is certainly warranted; if successful, they would have significant implications for early childhood education. Future cognitive research on the nature of the link between music, rhythm, spatial skills, and math would also be welcome.

The question of whether spatial ability, and presumably the math skills that are associated with it, can be taught takes us to the thorny issue of transfer—that is, the generalizability of skill from one area to another related one. One view of transfer holds that common neural pathways allow for the exercise of skill in one domain to stimulate activation in related domains, such as between music and spatial ability.[76] A contrasting view is that transfer is inherently conservative—that is, it is a fact of human nature that skill does not easily generalize, ensuring that the problem solver does not get stuck with irrelevant solutions.[77] Some experimental evidence suggests that it is easier to train a new skill by transfer than to change the manner of performing an old, well-practiced one.[78] Of course, as psychologist D.K. Detterman noted, "Both failure to transfer and appropriate transfer are equally important in major human accomplishment."[79] Most of education involves teaching a set of rules or concepts and the circumstances in which they apply (with frequent reminders of both).[80] Spatial skills, however, deal with the transfer of more fundamental mental abilities. Future studies should investigate whether individuals who are taught spatial skills can apply them to the same variety of tasks as individuals who already have successfully developed such skills. The following questions should also be considered in future research:

1. *What are the expectations for training?* Many studies examine results after a fixed period of practice or training; another (and perhaps more telling) method would be to train until performance levels off and ceases to improve. Few studies have tested the limits of training.[81]

2. *Whom does the training help?* Can it bring the performance of students with specific impairments in, say, mental rotation or route learning into the normal range? Can it teach spatial visualization to iconic visualizers or verbalizers?[82]

3. *How broadly can the training be generalized?* Does training improve performance on a specific spatial task, on a wide range of spatial tasks, or broadly on math problems that can be represented and solved by spatial imagery and manipulation?

4. *How long do the training effects last?* Are they ephemeral (i.e., a day, week, month) or do they bring long-term improvements?

5. *Does training in various spatial skills bring about a new way of looking at problems?* Do students gain an enhanced menu of strategy options, all easily accessible, from which to draw for solving a wide range of problems?[83]

Conclusion

One's intuitive sense of time, space, and motion lies at the heart of mathematics. From it, one derives by analogy a sense of number: the ability to compare, manipulate, and estimate quantities and to judge whether the results of exact calculations make sense. Moreover, spatial sense is essential to the grasp of basic mathematical concepts and relationships and the ability to solve problems.

Learning mathematics does not absolutely require spatial images and analogs. Students can succeed by mastering math facts, algorithms, and the internal logic of arithmetic, algebra, and geometric proof. But for many people, full understanding and greater efficiency depend on the ability to translate those conventions into a mental image. Indeed, the ability to move back and forth easily between the narrative or notational expression of a problem and a schematic mental picture of it lies at the heart of mathematical and scientific reasoning and characterizes those who are highly skilled in mathematics.

Most typically developing children come equipped, by virtue of some combination of innate cognitive capacity and informal education, with the intuitive underpinnings necessary to make sense of the basic conventional mathematics taught in school. With proper explicit instruction that links their intuitive sense of quantity, space, time, and motion with formal mathematical constructions, they can master the fundamentals. However, students who cannot envision and manipulate schematic pictures, as well as students who have lost the connection between pencil-and-paper conventions and the time-and-space events they represent, must draw on other resources—the linguistic and logical skills discussed in the next chapters. For some, this alternative path will be sufficient; for others, it may be insufficient and limit their mathematical education.

Notes

[1]Mathewson, 1999.

[2]Burnett, Lane, & Dratt, 1979; Friedman, 1995; Humphreys, Lubinski, & Yao, 1993.

[3]See, e.g., Mazzocco & Myers, 2003.

[4]See, e.g., Geary & DeSoto, 2001; Linn & Petersen, 1985; Masters & Sanders, 1993; Stanley, Benbow, Brody, Dauber, & Lupkowski, 1992; Voyer, Voyer, & Bryden, 1995.

[5]See Halpern et al., 2007.

[6]Hespos & Rochat, 1997; Örnkloo, 2007.

[7]See, e.g., Levine, Huttenlocher, Taylor, & Langrock, 1999; Nuttall, Casey, & Pezaris, 2005; Owen & Lynn, 1993.

[8]Opfer, Young, & Krasa, 2008.

[9]See, e.g., Corballis, 1997; Perrucci, Agnoli, & Albiero, 2003; Vandenberg & Kuse, 1978.

[10]Corballis, 1997; Pezaris & Casey, 1991.

[11]Hegarty, 2004, p. 280.

[12]Hegarty, 2004; Hodges, Spatt, & Patterson, 1999.

[13]See A. Lupkowski, as cited in Stanley et al., 1992, pp. 48–50.

[14]Spelke, Phillips, & Woodward, 1995.

[15]Gelman, Durgin, & Kaufman, 1995.

[16]Baillargeon, 2002.

[17]Case & Griffin, 1990; Siegler, 1976.

[18]Newcombe, Sluzenski, & Huttenlocher, 2005; Sluzenski, Newcombe, & Satlow, 2004.

[19]Postma, Izendoorn, & DeHaan, 1998.

[20]Cheng & Newcombe, 2005; Lee & Spelke, 2008.

[21]Hermer-Vazquez, Spelke, & Katsnelson, 1999.

[22]Moffat, Hampson, & Hatzipantelis, 1998.

[23]Newcombe & Huttenlocher, 2000.

[24]Dehaene, Izard, Pica, & Spelke, 2006; Delson, 2006.

[25]Beilstein & Wilson, 2000; Galea & Kimura, 1993; Saucier, Bowman, & Elias, 2003; Saucier et al., 2002.

[26]Quinn, Slater, Brown, & Hayes, 2001.

[27]Dehaene et al., 2006.

[28]Clements, 2004.

[29]Kulp, 1999.

[30]See Mix, Huttenlocher, & Levine, 2002b, pp. 63–80, for a review. See also Jeong, Levine, & Huttenlocher, 2007.

[31]Mix et al., 2002b.

[32]Mazzocco & Devlin, 2008.

[33]Cocude, Mellet, & Denis, 1999; Kosslyn, 1994; Ungerleider, 1995.

[34]Kozhevnikov, Hegarty, & Mayer, 2002.

[35]Cheng & Newcombe, 2005.

[36]Harris & Miniussi, 2003; Jordan, Wüstenberg, Heinze, Peters, & Jancke, 2002; Marshall & Fink, 2001.

[37]Ungerleider, 1995.

[38]Perrucci, Agnoli, & Albiero, 2008.

[39]Jordan et al., 2002. See also Ark, 2005; Kucian, Loenneker, Dietrich, Martin, & von Aster, 2005.

[40]Corballis, 1997; Franco & Sperry, 1977.

[41]Wendt & Risberg, 1994.

[42]Kozhevnikov, Kosslyn, & Shephard, 2005.

[43]Kozhevnikov et al., 2002; Kozhevnikov et al., 2005.

[44]Sternberg & Weil, 1980.

[45]Lowrie, 2000.

[46]Blajenkova, Kozhevnikov, & Motes, 2006; Kozhevnikov et al., 2005.

[47]Hembree, 1992; Kozhevnikov et al., 2002; Sternberg & Weil, 1980.

[48]Knauff & Johnson-Laird, 2002; Kozhevnikov et al., 2005.

[49]McGuinness, 1993, p. 264.

[50]Hegarty & Kozhevnikov, 1999, p. 686.

[51]See Arcavi, 2003; Presmeg, 1986.

[52]Wolfgang, Stannard, & Jones, 2001.

[53]Webb, Lubinski, & Benbow, 2007.

[54]Hermelin & O'Connor, 1986; Lehmann & Jüling, 2002; Stevenson et al., 1985.

[55]MacFarland Smith, as cited in Habre, 2001, p. 43.

[56]Tartre, 1990.

[57]Gallagher, 1992.

[58]Burnett et al., 1979; Casey, Nuttall, & Pezaris, 1997; College Board, 2007.

[59]Gallagher, Levin, & Cahalan, 2002; see also Johnson, 1984.

[60]Casey, Nuttall, & Pezaris, 2001, p. 57. Math problem: Reprinted with permission from Journal of Research in Mathematics Education copyright 2001 by the National Council of Teachers of Mathematics. All rights reserved.

[61]Casey et al., 2001.

[62]Burnett et al., 1979.

[63]Royer & Garofoli, 2005.

[64]Low & Over, 1993.

[65]Lim, 1994; Meehan, 1984.

[66]McGuinness, 1993, p. 266.

[67]Tartre, 1990.

[68]Shea, Lubinski, & Benbow, 2001.

[69]See, e.g., Jordan, Levine, & Huttenlocher, 1995; Shafrir & Siegel, 1994.

[70]Jordan, Hanich, & Kaplan, 2003b. See also Ostad, 2000; Robinson, Menchetti, & Torgesen, 2002.

[71]Liu, Uttal, Marulis, & Newcombe, 2008; Newcombe, Mathason, & Terlecki, 2002; Terlecki, Newcombe, & Little, 2008.

[72]Ferrini-Mundy, 1987; Piburn et al., 2005; Sorby & Baartmans, 2000.

[73]Casey et al., 2001.

[74]Rauscher, 2003.

[75]Graziano, Peterson, & Shaw, 1999.

[76]Leng & Shaw, 1991.

[77]Detterman, 1993.

[78]Opfer & Thompson, 2008.

[79]Detterman, 1993, p. 4.

[80]Barnett & Ceci, 2002.

[81]Newcombe et al., 2002.

[82]Kozhevnikov et al., 2002.

[83]Kozhevnikov et al., 2005.

The Language of Mathematics

N umber sense and good spatial ability allow people to grasp the essential meaning of number and mathematical relationships, make quick estimates of quantities and calculations, check the reasonableness of computations, and solve novel problems efficiently. For anyone who pays taxes, builds bridges, or pilots airplanes, however, those skills—although important—are not sufficient. In these situations, people must also be exact. Exact mathematics requires counting; reading, writing, calculating, and solving problems with Arabic notation; expressing ideas in algebraic formulas and graphs; and so forth—in short, being fluent in the language of conventional mathematics.

CALVIN AND HOBBES © 1988 Watterson. Dist. by UNIVERSAL PRESS SYNDICATE. Reprinted with permission. All rights reserved.

Unless one is lucky enough to be a stuffed tiger, one must be taught this language. Conventional mathematics, unlike the informal variety, is not intuitive—indeed, it is often counterintuitive—and requires a great deal of instruction, practice, and hard work. Although some educators refer to math as a "second language," children really learn mathematical language as another "first language" right along with their natural, conversational tongue at home and school, beginning with counting.[1] As Section II reviews the cognitive and neuropsychological underpinnings of learning and doing conventional mathematics, we examine the ways those mechanisms can go awry and, when they do, how such failures affect mathematics learning.

Chapters 4 and 5 explore the challenges of learning the spoken and written forms, respectively, of mathematical language. In doing so, these chapters take up the thorny question of the relationship between mathematical learning impairments and disorders of language and reading.

Chapter 6 examines the research on brain function and mathematical language. The neuroscience on natural language and reading is well ahead of that in mathematics, but advances in this field are rapidly closing the gap. This chapter discusses several lines of investigation that should provide useful insights for the classroom teacher.

Finally, in Chapter 7, we discuss a series of issues and topics in conventional mathematics that have proven particularly vexing for students, including the vocabulary of the mathematics classroom, word problems, bilingual math education, notational and conceptual directional orientation, and fractions.

Note

[1]Wakefield, 2000. See also Pimm, 1987.

Speaking Mathematics

Mathematical precision begins with enumeration. How do children learn to count? Considerable scholarly controversy exists regarding this feat, centered on the question of infants' quantitative skills.[1] The conservative position—the one adopted herein—holds that counting results from the tightly intertwined development of both new quantitative concepts and new linguistic skills in toddlers and preschoolers. Chapter 4 begins with a discussion of counting and then introduces some of the basic mathematical concepts that counting makes possible.

Counting

Counting is not universal: Some less advanced societies have words for only "many," "few," and "fewer."[2] Counting—attaching number names to objects according to a set of rules—is a cultural achievement.[3] Although it sounds simple and is usually mastered easily with help from home and school, counting is actually a complex skill with several underlying components. The discussion that follows reviews its development in young children.

Tracking and Distributing

How infants think about quantity is not fully understood; however, the fact that infants possess important underlying capabilities is well established. For example, one of the earliest things infants know about objects is their location in the environment. Research has shown that beginning in the first 6 months of life, infants can simultaneously locate, track, and anticipate the trajectory of two or three objects as those objects move, even when the objects pass out of view or when other features of the objects (e.g., shape, color) change.[4]

By the middle of the second year of life, children can not only visually track objects, but can also handle and manipulate them purposefully, such as when they give, take, or relocate a toy. Fueled by normal social incentives, informal instruction, and encouragement, toddlers become increasingly systematic in this behavior: distributing cookies at preschool snack time or putting toys in their proper places. These sequential, one-to-one matchup experiences serve as a prelude to counting. For example, such activities demonstrate to toddlers that it does not matter in what order you hand out the cookies as long as every classmate gets one—and that to be fair, no one should get more than one. Children first understand these pairing rules not in terms of numbers (i.e., four cookies for four classmates), but on the basis of one handout at a time. A child's skill at this point depends on the context. That is, children can demonstrate complete competence in some settings (e.g., snacks) and none at all in others (e.g., cleanup)—social and emotional salience is key. Children learn the rules of fair distribution through guided experience.[5]

What kinds of things can be paired? In these early years, children learn to distinguish "things" from "stuff." Although infants can track things (e.g., blocks), they do not track stuff (e.g., sand). This distinction becomes especially relevant and reinforced as toddlers learn that they can more easily and precisely manipulate and distribute discrete objects, such as dolls or cookies, than they can nondiscrete material, such as clay, sand, or applesauce. Although they can identify "more" and "less" applesauce, it is difficult for them to dish it out fairly, whereas cookies usually come in distinct units ready for equitable distribution. If there are not enough cookies to go around, children younger than 3 will typically break up what they have so that everybody gets a piece, regardless of the size of the pieces. They understand that fairness pertains to numerosity and that numerosity is unrelated to the size of the objects.[6]

As previously discussed, the quantity-sensitive areas of the brain are embedded in a broad region that also allocates attention, tracks objects through time and space, and maintains hand and finger awareness and control—all skills that young children use when they point to and count objects. Fingers are especially linked to number and counting. For example, pairing fingers to objects, as well as to number names, is a nearly universal step in the process of learning to count; moreover, the most widely used number system is based on 10, and the Latin word *digit* applies to both fingers and numerals. Several lines of research support the connection between fingers and numbers. For example, researchers covered first-grade students' hands and touched several fingers in turn; the children's ability to identify which fingers had been touched when the cover was removed predicted their math ability a year later. Even among 10- to 12-year-old students, finger awareness was associated with performance on simple calculations.[7]

The observation that blind children do not use their fingers for counting suggests that fingers may convey their important quantitative information visually rather than through touch or movement.[8] One may speculate that fingers, as they are naturally lined up on the hand, serve as children's first number line, allowing them to attach the ideas of "more" and "less" to counting numbers. The importance of fingers raises the concern that an impairment in finger awareness or control may compromise a child's acquisition of counting skills. In fact, early counting skills have been linked to degree of finger control in 3-year-old children with spina bifida.[9]

Early Number Language

True counting begins as children learn to talk. This section examines how children learn the number sequence and figure out what to do with it, as well as other linguistic contributions to early number concepts, such as location words and the grammar of big numbers.

Number Words

Number words emerge in speech around the end of the second year, often beginning with *two*, probably not coincidentally around the second birthday, when children are taught how to respond to the question, "How old are you?" (although they probably do not know the meaning of *two* or *old*, for that matter). *Two* is also the smallest necessary counting word, because in English a single item is typically referred to by the article *a* or *the*. Children learn number words in the same way they learn other words: from ambient conversation (about 40% of mothers' number-word use is incidental) and informal instruction.[10] Although children use number words imprecisely at first, in a manner that is disembodied from actual counting, they are usually referring to numerosity. That is, a child might say "three" to refer to four things, but nevertheless, the child is talking about how many items there are.[11]

By their second birthday, toddlers have also figured out the difference between *is* and *are*. When children are told, "Look, there are some toys," they expect to see more than one. By age 3 years, children have typically picked up the basic spoken lexicon and syntactic rules of number. For example, they learn that plural nouns (i.e., *kitties*) refer to more than one countable thing. They also learn that number words and other determiners (e.g., *a, some, both, another, all*) are special descriptors that refer to quantity, come first in a series of descriptors (i.e., *three red apples, a very good dog*), and often refer to a group of things. Furthermore, preschoolers understand that unlike other determiners, number words refer to specific, unique numerosities, even if they do not yet know what those are; that is, *six* does not just mean *a lot* the way *some* can. Thus, when a collection's numerosity changes, it needs to be described by a different number word.[12] Number knowledge begins with the smallest quantities and gradually builds up. Even though children know *a million* is a specific amount, it may simply connote *incomprehensibly large* for many years, as discussed in Chapter 2 in relation to the mental number line.[13]

Children are especially sensitive to the sound patterns of number words as they learn to count.[14] To understand spoken number words, children must be able to distinguish reliably between the sounds that make up the word and to perceive the sounds in a timely fashion. Adequate auditory processing is important for discriminating between similar-sounding number words like *thirty* and *thirteen*, for example, or among words that are rattled off rapidly or in noisy surroundings.[15] To understand verbal expressions, children must be able to perceive and remember sounds in order. This latter skill is especially important to become fluent in the counting-word sequence; children with phonological processing impairments are often slow to master counting. The ability to perceive and remember sounds in order may also play an important role later in mastering arithmetic facts and oral word problems. In fact, a study of first graders found that phonological skills were significantly linked to arithmetic fact fluency.[16]

Early Counting

In the same informal and uneven manner that young children learn most of their math skills, they learn four important characteristics of number words and how to use them. They are listed sequentially, but students learn them concurrently and often independently of one another.[17] First, children learn that number words have a fixed order; indeed, they often learn the sequence as one word: *onetwofreefourfive!*[18] When children start learning the number-word sequence early in their third year, they begin to pay more attention to the countability of things, further enabling them to distinguish cookies from applesauce.

Second, children learn that number words can be paired with objects and events, just as cookies can be matched to friends at snack time. This verbal tagging usually involves coordinating the verbal sequence with some sort of action, such as pointing to fingers or objects, and follows the familiar fair-share rules: one and only one unique number word for every object. Through experience, increasing linguistic competence, and informal education, children learn to tag objects with words in sequence beginning with *one*. [19]

Children also learn that, just as they can hand out cookies in any order at snack time (as long as everybody gets one), they can also point to the objects in any order. Although children do not seem to have a problem with arbitrary order when they hand out cookies, coordinating unsystematic pointing with an ordered word sequence is sometimes very confusing. For this reason, children are frequently taught to count using objects that are lined up; indeed, the most commonly counted objects, fingers, come that way naturally. Thus, young students often do not figure out that counting things in a certain order, such as consecutively from left to right, is merely a convention and not a rule until long after they have become quick and accurate counters; many children who have math difficulties never learn that principle. [20] Furthermore, the insight that an object can be tagged with a number word regardless of the object's location is significant: It is linked to the emerging idea (known as "conservation of number") that a collection's quantity does not change when the objects are moved around.

Another cog in the mental counting apparatus is the realization that the number word paired with the object counted last in a collection stands for how many objects are in it—the collection's "cardinal value." This recognition emerges gradually by about age 4 through counting activities. Linguistically speaking, cardinality is a collective descriptor. Except for *one,* it necessarily describes an attribute of a set of multiple things rather than an individual item, and it is learned at this age in tandem with mastering general collective vocabulary and syntax (e.g., *crowded class*). Just as learning the word *red,* for example, heightens children's awareness of color and gives them a way to answer the question, "What color?", so counting sparks their interest in the cardinality—the "three-ness," for example—of collections and gives them a way to answer the question, "How many?" [21] Using the number-word sequence to find out how many objects are in a collection leads to the idea that each successive number represents one more object than the number before it—a notion known as the "$N + 1$ rule." In this manner, counting enhances children's grasp of relative value.

By some time in their sixth year, children have typically learned all the basic principles of counting, although it can take years for students to become proficient in applying them. (Typical 5-year-old kindergartners can count at least six objects, but only about half of those children can determine when two sets are equivalent, for example. [22]) Counting also includes the notion that any set of discrete objects or events can be counted—that is, counting is abstract and the objects' features are irrelevant (see Chapter 9). When children understand that the last number word in a counting sequence tells "how many" there are, they have begun to emerge from the fuzzy, distorted, and limited world of approximate numerosity into the focused world of exact and potentially limitless enumeration.

Location Words: The Language of Space

Just as the counting words give voice to numerosity, location words and phrases express spatial relationships. Children as young as 2 years begin to grasp prepositions and prepositional phrases, deciphering the code from syntax and context. Pre-

BABY BLUES © BABY BLUES PARTNERSHIP. KING FEATURES SYNDICATE. Reprinted by permission.

schoolers also learn other positional descriptors. For example, when they line up, they learn what it means to be "first," "last," or "next" in line. During the early years, children learn the vocabulary and syntax of spatial description, heightening their awareness of linear, graduated relationships.[23]

Spatial language supports the notion of ordered values. Words and concepts such as *before, after, behind,* and *in front of* draw attention to sequencing, a step toward establishing a mental number line. Spatial language also allows children to master geometric concepts—to analyze, remember, and communicate angles, distances, directions, landmarks, shapes, and their relationships.[24] Although there are similarities between linguistic and spatial (i.e., drawn, gestured, or modeled) communications, the two are fundamentally different in that language is necessarily sequential: One word follows another.[25] Language that describes spatial relationships can thus enhance one's ability to analyze problems with elements that are perceived concurrently.

Number Grammar: Big Numbers

In English, learning the number words *one* to *ten* is straightforward; numbers larger than *ten*, however, are trickier. Learning the words for big numbers ultimately involves mastering the base-10 numbering system, a task aided by mathematical notation (see Chapter 5). This section focuses on the verbal demands and difficulties inherent in big numbers.

Mathematical language, like natural language, has a grammar. For example, through *ten*, numbers are considered *lexical* constructs—like simple words (*six, flower*). Numbers greater than *ten* are *morphological* constructs: They express base-10 place value by linking number words together additively, like compound words (*twenty-one, birthday*). Mathematical operation statements are *syntactic* constructs, like sentences (*one plus two equals three, I went to school*).[26] All of these constructs carry meaning, or *semantic* value; in the case of a number, the semantic value is the quantity it signifies. This grammatical categorization applies alike to number words and symbolic numerical expressions.

Unlike Arabic numerical notation for big numbers, which is now used almost universally, the verbal expressions for quantity retain closer ties to local cultures and languages; like other localisms, their meanings easily get lost in translation. The term for 94, for example, is *ninety-four* in English, and translates to *four and ninety* in German and Dutch and *four-twenty-fourteen* in French—the latter construct a vexing holdover from an archaic base-20 system.

st to Western languages, Chinese and other related East Asian lan-
everal useful linguistic features that make them especially suited for
,.[27] These characteristics have been linked to East Asian students' supe-
chievement as early as entry into first grade. Strikingly, Chinese number-
hology is completely congruent with the base-10 concepts that govern
re of the numerical symbols. For example, *twelve* translates to *one-ten-two*
fifty-three is rendered as *five-ten-three*. In contrast, English-speaking children must
conquer an irregular and archaic math lexicon: Words such as *eleven, twelve, twenty,*
and *thirty,* as well as the morphological reversals in the teens, must be mastered to
count above 10. In fact, any English speaker wishing to count to 1 million must memorize 30 different word names, whereas a Chinese speaker needs to know only 14.

Because Chinese dispenses with the grammatical distinction between singular
and plural, incorporating cardinal numbers in speech and writing is much easier
than it is in English, in which someone would say "one bird flies" but "two birds fly."
In addition, while English has three words and one suffix signifying ordinality (*first,
second, third,* and *-th*), Chinese simply adds the prefix *dì*. Chinese-speaking children
therefore master both large-number counting and base-10 concepts much earlier
than their English-speaking counterparts, at least in part because they have fewer
linguistic hurdles. Although one may think of counting as a merely preliminary
phase of mathematics education, among English speakers counting mastery is robustly associated with math achievement even as late as the seventh grade—well past
the time when the curriculum focuses on that skill.[28]

Early Arithmetic

During the fourth year of life, counting opens the door to simple arithmetic.[29] For
example, one of the things counting allows children to do for the first time is to compare quantities precisely and with confidence: The plate with five cookies on it def-
initely has more cookies than the plate with three. Counting allows children to say
how many more—in this case, two more cookies. Older children can solve this problem abstractly by subtraction; without that sophisticated tool, younger children
tackle this sort of problem concretely by counting. For example, researchers showed
a picture of five birds and three worms to preschoolers and asked them, "Suppose
the birds all race over and each one tries to get a worm. How many birds won't get
a worm?" Nearly all the children answered correctly by first counting out the worms,
then counting out the same number of birds, and then counting how many birds remained.[30] Children can handle subtraction's take-away function by counting as well.

Counting also opens the door to addition. Children normally learn to add by
using several strategies, all variations on counting. Typically, the earliest method is
counting all of the items to be added in a single sequence using objects, fingers,
and/or number words. A more sophisticated approach—requiring the ability to
compare values—is the minimum addend strategy (see Chapter 2). Once children
have mastered arithmetic facts, they often find that retrieval of the information
from memory is the most efficient approach, although children's arithmetic strategies do not mature in as straightforward or thorough a manner as was once thought.
Children who know several strategies generally choose one based on a cost-benefit
analysis of how quickly and/or accurately the method will let them solve the particular problem at hand.

Communication Disorders
and Mathematical Language

Under ordinary circumstances, children develop number-word skills unimpeded. Many children grow up under more challenging conditions, however, and experience considerable difficulty in mastering the language necessary for counting. This section considers three special cases: deaf children, children with developmental language disorders, and children with linguistic deficiencies due to poverty and cultural deprivation.

Research on young children with severe hearing impairment, which makes adequate phonological input impossible, reveals they have trouble communicating concepts and learning the counting words. Furthermore, children who use sign language often become confused when they cannot simultaneously use their fingers for counting; consequently, they understand and communicate mathematical ideas more effectively through activity and objects rather than through language. For the most part, however, deaf children and adults demonstrate normal number sense. They can compare quantities easily and their comparisons reveal the same ratio effects seen in those of people without hearing impairments, although deaf students often make those judgments less efficiently. Overall, the mathematical development of children with impaired hearing parallels at a slower rate that of students with normal hearing.[31]

A problem with number words + + + + + + + + + + + + +

Annie is a soft-spoken 8½-year-old girl who talks with immature articulation. She was brought to our attention because she lags far behind her second-grade classmates in learning how to count.

Annie's birth and early development were unremarkable except for severe articulation difficulties. She is now also having trouble learning to read and spell. When evaluated at age 6, her verbal and nonverbal skills were highly discrepant, measuring around the 20th and 90th percentiles, respectively, and her phonemic awareness was poor. On her sixth birthday, she was unable to answer the question, "How old are you today?" without looking at her fingers for a visual cue. Although her verbal skills improved dramatically over the ensuing year and a half, she did not master the alphabet until she was almost 8.

Through the first half of second grade, Annie did all her arithmetic on her fingers. Since then, however, she has caught on quickly to addition strategies, such as recomposing 9 + 7 to 10 + 6, and is currently ahead of her class in those concepts. She has no difficulty picking the larger of two numbers up to 100, either orally or in writing, although when asked orally for the larger of 22 and 17, she misheard "seventeen" as "seventy" and answered accordingly. She has no trouble ordering three-digit numerals.

Where Annie does struggle, however, is with counting and reading numerals aloud. The table shows her responses to oral and written questions. As one can see, she makes lexical and sequencing errors. She also makes reading mistakes, a topic covered in Chapter 5.

Questions/requests	Annie's responses
"What is six plus six?"	"a one and a two"
"Count from seventy-five to ninety-three."	"seventy-five, eighty-five, ninety-five, eighty-one, eighty-two, eighty-three . . ."

Questions/requests	Annie's responses
"Count backward from twenty-two."	"twenty-two, twenty-two, twenty-two, twenty-one, ten, nine, eight, seven . . ."
"What comes just before thirteen?"	"thirty-two"
"What comes just before sixteen?"	"seventeen"
"What comes just before twenty-one?"	"I don't know."
"What number comes just before nineteen?"	"one and eight"
"Just after nineteen?"	"two and zero"
"Fill in the missing numbers: ___ 19___"	18, 20
"Read the answers you just wrote."	"twenty-eight and thirty"
"Read: 250"	"two hundred five"
"Read: 15"	"twenty-five"
"Read: 12"	"twenty-two"
"Read: 21, 37, 82"	"twenty-one, thirty-seven, thirty-eight"

On formal evaluation near the end of second grade, Annie scored below the 15th percentile for her age on tests of rapid object and numeral naming, suggesting that her difficulty attaching number names to quantities and written numerals may reflect general naming difficulties. She also performed slowly on a test of numeral matching, where being able to attach names to the numerals would have helped her match them more efficiently.

Annie has excellent number sense and understands the quantities represented by the written digits. Her teachers, however, wonder how far she will be able to advance in her mathematical conceptual development without first mastering the number words.

Unlike hearing impairment, developmental language disorders are characterized by impairments located in the brain, not the ears, and often have a more significant impact on children's mathematical abilities. One study found that children with developmental language disorders manifest severe math impairment at about 10 times the general population rate by the time they reach the intermediate grades.[32] Among many children with language delays and severe phonological difficulties, math skills are at least as compromised as reading ability, and these children's counting skills are frequently delayed. When they do count, they do so slowly to maintain accuracy and often fail to become fluent with the counting sequence. Memorizing arithmetic facts is exceptionally difficult for them as well. Some investigators attribute these counting difficulties to impairments in articulation, expressive sequencing, and (among those children with both receptive and expressive delays) phonological sequencing and verbal memory. Other scholars, however, speculate that these children's language and math disorders are independent and reflect underlying brain dysfunctions affecting neighboring brain regions. [33]

Slow to speak +

Keith, a 9-year-old third grader, demonstrates the complex role that language plays in mathematical learning. Born prematurely after a difficult gestation with poor prenatal care, Keith was adopted into a warm, supportive, and well-educated family. Despite his advantages at home, his language development lagged about 6 months behind that of his peers. His parents first became concerned about Keith when he could not rhyme words as a toddler, and weak phonemic awareness has contributed to his reading difficulties.

The most striking aspect of Keith's language, however, is that he is exceptionally slow to respond. Although he understands everything that is said or read to him and has good knowledge of word meanings, his ability to express himself adequately and in a timely manner—to name common objects, formulate sentences, respond in class, and even converse with classmates—is painfully compromised. Even visual tasks, such as identifying related pictures, go more slowly for him than for others his age because he does not have ready access to the words that would help him solve the problems more efficiently.

The effect of this difficulty on Keith's math is that he is very slow on timed arithmetic fact tests. On a timed test of simple addition, subtraction, and multiplication facts, he performed around the 10th percentile for his age. During class, his teachers and classmates know that when Keith is called on for an answer, they will have to wait a long time before he speaks.

His teachers and classmates also know that it would be a mistake to confuse Keith's tardiness with lack of math ability. In fact, Keith is regarded as "smart" when it comes to math. He prefers to compute in his head, possibly because it is even more laborious for him to write the problems out than to do them mentally; his answers, however, are almost always correct. For example, when asked aloud for the product of six times nine, after several minutes he said, "fifty-four." When asked if he had remembered that information or had figured it out some other way, he responded slowly as follows: "I know that nine plus nine is eighteen. So you add eighteen plus eighteen plus eighteen. First I added ten plus ten plus ten. That's thirty. Then I added eight plus eight; that's sixteen. So I added another ten to the thirty to get forty. Then I added six to eight and got fourteen. And then if you add fourteen to forty, you get fifty-four."

The composite of Keith's number placements on the 0–100 number line demonstrates his age-appropriate mental number line.

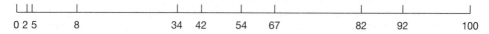

| | | | | | | | | | | |
|0|2 5|8| |34|42| |54|67| |82|92| |100|

Keith is likewise sensitive to number patterns and can infer missing elements in number series. Moreover, he loves to build things and has superior visual-spatial abilities; his performance on a mental rotation test measured in the 99th percentile for his age. This means that when Keith cannot marshal the words necessary to solve a problem, diagrams and models may benefit him enormously.

Some children with language disorders understand counting principles such as cardinality. Among children with intact nonverbal skills, the ability to compare small numerosities based on visual cues often develops normally, providing further evidence that number sense is partially independent of language. Among the few reported cases of children with language impairments who had attained average-range math abilities by adolescence, all began with intact nonverbal skills and had overcome their early language difficulties—highlighting the importance of both verbal and nonverbal skills for mathematical development. However, most children with language disorders are not able to compensate adequately; their math skills decline even further by high school in relation to both their own reading levels and the math skills of adolescents without disabilities.[34]

Verbal sequencing +

Jessica is a seventh grader who was brought to our attention because of her poor number sense. Serious medical difficulties during gestation and her early years left Jessica somewhat inattentive, slow, and forgetful, which has made new learning difficult. Her early speech development was characterized by poor articulation, as well as by grammar and vocabulary limitations; at age 6, she was found to have an auditory discrimination dysfunction that affects her reading and spelling.

The most striking aspect of Jessica's language and thinking when we interviewed her at age 13 was her apparent lack of fluency in the kinds of overlearned verbal sequences that most people know automatically. For example, when asked how many seasons there are in a year, she confused them with the months, even when given a clue. When then asked how many months there are in a year, she recited them under her breath slowly and with much effort, arriving at an answer of "ten." On a second interview, Jessica recited them aloud, successfully getting to July, eventually adding August, and only after much thought, produced the remaining months. She then wondered aloud, "Oh, March! Did I say March?" Jessica could recite the days of the week only by singing them as a song, saying, "That's how I remember all of it." Lacking automatic mastery of these sequences, she sometimes has difficulty orienting herself when starting anywhere other than the beginning. For example, although she could recite the alphabet, when asked whether *D* or *K* is closer to *F*, she chose *K*.

Jessica has similar difficulties with number sequences. She must still rely on the song she learned in third grade to remember her multiplication facts and uses counting or unreliable recall for the other operations. On a timed test of mixed simple arithmetic problems, she scored in the 8th percentile for her age, for the most part because she worked very slowly. Although Jessica can count forward, backward, and by twos, it is not always automatic. For example, when asked what number comes just before 67, she had to think before arriving at "sixty-six." When asked to count from 75 to 93, she went from "seventy-nine" directly to "ninety" before realizing she had left out the 80s. Without an automatic sequence on the tip of her tongue, she must rely on working memory to a greater extent than other students.

Furthermore, Jessica's number-line placements are immature. When asked to locate 100 on a 0–1,000 number line, she oriented herself by marking where she thought 10 should be, as shown in the composite drawing of her placements.

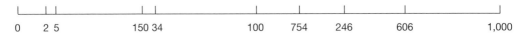

0	2 5	150 34	100	754	246	606	1,000

As is typical of much younger children, Jessica has an overblown idea of the relative value of small numbers and only vague notions about the larger ones. For example, when asked to choose quickly which number, 89 or 73, is closer in value to 78, she chose 89.

Lacking a serviceable number–word sequence or mental number line, number patterns elude Jessica, and she is left with only effortful counting and algorithms at her disposal for solving problems. For example, when asked to mentally subtract 3 serially from 35, she "wrote" each algorithm on the table with her finger as she spoke, leaving her vulnerable to error: "thirty-two, twenty-eight, twenty-five, twenty-two, eleven, eight. . . ." Given the sequence 23, 26, 30, 35, ___, Jessica wrote down a list of all the numbers between the given ones instead of reciting them in her head or subtracting.

Jessica's situation is obviously complicated. It is difficult to determine with certainty which set of difficulties—memory, attention, language, or number sense—is primary. Nevertheless, it is clear that without verbal sequence fluency, including numbers and number facts, a student is left to reinvent the wheel with every problem.

Because basic n   —the earliest linguistic math skills—
are cultural pher   endent on environmental input at
home and presc   cultural deprivations. Investigators
found that the   5-year-old children from low-income
families in the   nd 4-year-old children from average-
income families. Their counting skills were poorly developed and they made little
use of their fingers. These children began their addition counting procedure either
with the smaller addend instead of the larger one or with some number other than
either addend—an approach rarely seen in more advantaged children. Moreover,
difficulties in basic linguistic skills, such as vocabulary and syntax, impaired their
comprehension of story problems and verbal procedural cues.[35]

Classroom Implications

Math education begins during toddlerhood and the preschool years, when opportunities for distributive activities abound. Fun and meaningful tasks that pair objects
one to one will establish a firm foundation for counting, as will learning the counting sequence. It is also possible that finger games and finger counting would foster
finger awareness and introduce ordinality, an approach that may prove crucial for
children from low-income households.[36] Like most methods we report, however,
these arc untested and merit research attention.

Teachers have at their disposal many ways to teach counting, and typical preschool and elementary classrooms are usually well stocked with beans and buttons.
One of the toughest challenges for some young children, however, is coordinating
pointing with saying the number words. Frequently, they understand the matching
concept but lack the physical coordination. One unproven pedagogical technique
is to instruct one component at a time, with the teacher pointing or counting for the
student as needed; in theory, children can develop the speech and movement components separately and put them together when they are ready.

Children's key task at this early age is to link the number words to their relative
values, so that they learn "five" means more than "three." Counting can be directly
tied to the number line through the many number-sense enhancing activities discussed previously: playing linear numerical board games and hopscotch; counting
off on the lunch line, stairs, and monkey bars; and learning the musical keyboard
and rhythm. Furthermore, complementary movement and verbal activity should enhance coordination in physical counting. Teaching counting only in this orderly
array will be incomplete, however, if it neglects the concept that items can be
counted in any order. For example, Jenny, José, and Jacob can line up for recess in
any order, but the first person is always "one." Teachers can easily recruit such
simple, entertaining activities to foster insight into the relation between the verbal
counting sequence and the number line.

A related untested but promising counting activity involves number blocks and
a counting board (Figure 4.1).[37] (The "manipulative" version is shown here, but
comparable software versions are equally promising.) In this task, children count
out unit cubes as they fit them into a graduated series of grooves in a board that holds
from 1 to 10 blocks. The counting board gives students hands-on experience with all
of the counting principles. Children attach a number word, drawn consecutively
from the verbal sequence, to each cube as they pick it up. They can pick up, name,

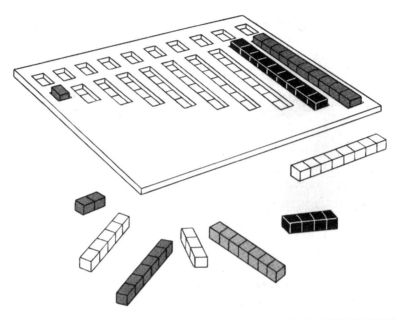

Figure 4.1. Counting board. (From CHILDREN DISCOVER ARITHMETIC: AN INTRODUCTION TO
STRUCTURAL ARITHMETIC, 2ND ED. by CATHERINE STERN and MARGARET B. STERN. Copy-
right © 1949, 1971 by Harper & Row, Publishers, Inc. Reprinted by permission of HarperCollins
Publishers.)

and use the cubes in any order and get immediate visual feedback about the number of cubes in a filled groove. Once children have a basic grasp of counting, they
can use notched number blocks, as shown in the figure. Children can also see from
the completed counting board's stair-step array that each consecutive groove holds
one more cube than the groove before it. Later, children can tag each column with
a numeral.

The counting board links numbers to the mental number line in several ways.
Vertically, each column demonstrates numbers as different lengths, like distances
along a number line. Horizontally, the columns of blocks represent consecutive cardinal values, lined up from small to big. In this manner, children learn about the
relationship between cardinality and ordinality, or how values can be ordered by
size. Cardinal and ordinal number words and number blocks together illustrate how
values are ordered: "One" cube lives in the "first" groove, "two" cubes live in the "second" groove, and so forth. Unfortunately, the pedagogical effectiveness of the
counting board—although a popular teaching device several decades ago, particularly in Europe—has never been subjected to scientific testing. Given the recently
demonstrated success of such number-line related techniques as the musical keyboard and certain types of board games, however, research on the counting board's
effectiveness is clearly warranted.

As with counting, instructors have many methods at their disposal for teaching
informal arithmetic. Most of these methods are extensions of the pedagogical devices used to teach counting: Number lines, linear numerical board games, and musical keyboards all offer opportunities to compare, add, and take away. By using complementary number blocks to fill an $N \times N$ frame (as shown in Figure 4.2 for $N = 3$
and $N = 4$), children learn all the combinations of two numbers that are "the same
as" N. They also see that in most cases each pair has two versions (e.g., "one and
three is the same as four, and so is three and one")—an important step to under-

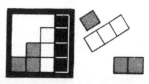

Figure 4.2. Using complementary number blocks to sum to 3 and 4. (From CHILDREN DISCOVER ARITHMETIC: AN INTRODUCTION TO STRUCTURAL ARITHMETIC, 2ND ED. by CATHERINE STERN and MARGARET B. STERN. Copyright © 1949, 1971 by Harper & Row, Publishers, Inc. Reprinted by permission of HarperCollins Publishers.)

standing addition's commutative property. Likewise, children learn that the N block fills its own column without needing another block (e.g., "four and nothing is the same as four"), which is a good introduction to the concept of zero. Finally, they see what remains when a block is removed from each column, a transformation that they will learn to model by subtraction. To find the difference between two values— another transformation modeled by subtraction—children can place the blocks side by side or one on top of the other; the unshared length is their difference. Again, none of these methods, including software versions, have yet been subjected to adequate scientific scrutiny; they all show promise, however, and their effectiveness warrants scientific investigation.

The *-teen* number words are particularly challenging, but can be demonstrated with the 10-block and unit cubes (Figure 4.3), providing a good introduction to the idea that the base-10 system is founded on counting and addition. Unfortunately, forcing base-10–friendly Chinese number words into the English lexicon is not a realistic option. One study showed, however, that with some classroom adjustments

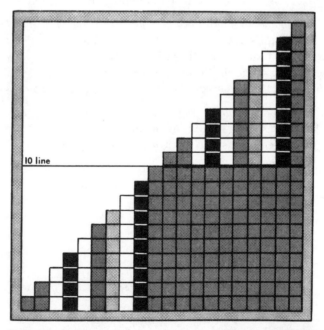

Figure 4.3. Learning the "teens" on a number frame. (From CHILDREN DISCOVER ARITHMETIC: AN INTRODUCTION TO STRUCTURAL ARITHMETIC, 2ND ED. by CATHERINE STERN and MARGARET B. STERN. Copyright © 1949, 1971 by Harper & Row, Publishers, Inc. Reprinted by permission of HarperCollins Publishers.)

in number language and explicit instruction in base-10 concepts, even U.S. first graders at risk for math difficulties learned big-number concepts easily.[38]

Research has demonstrated that the number-line counting activities described in Chapter 2 are effective in enhancing early numeracy among children from low-income homes. Many of the activities noted in this chapter also may be beneficial to them. Further research is warranted to determine what, if any, methods are effective for teaching counting to children with primary hearing or language deficits.

Conclusion

A child's first lesson in mathematical language is learning how to count—the bedrock of conventional mathematics. Like any linguistic task, counting depends on solid verbal skills learned informally in a language-rich environment, as well as on adequate auditory processing and speech. Learning the rules governing counting emerges from experience gleaned in myriad matchup activities at home and in preschool, and requires the ability to fluently integrate tracking, pointing, and speaking. Counting gives children two important pieces of numerical information. One is the exact numerosity—the cardinality—of a set of objects. The other is a more precise idea of relative value, or ordinality, the foundation of number sense.

Future research in early mathematics education needs to focus on three areas:

1. *How should counting be taught in preschool?* One of the greatest pedagogical challenges is keeping number sense alive as a child masters conventional mathematics. Thus, ordinality deserves special attention in curricular research. It will be especially important, of course, to gauge the effect of curricular and pedagogical choices on children's future mathematical achievement.

2. *How can children's cognitive readiness for learning mathematical language be assessed?* Children require intact phonological functions, receptive and expressive language, body (particularly finger) awareness, and the ability to integrate all of these. Future research should discern what skill level is required for early math learning and how to screen young children for impairments in these skills.

3. *What can we do for children who are not cognitively or perceptually ready to learn counting?* Future research should investigate whether impairments can be remedied directly or if compensatory approaches are required.

This chapter discussed "conversational" math, particularly counting, which is the foundation of conventional mathematics. Chapter 5 examines reading and writing mathematical language.

Notes

[1]See, e.g., Gelman & Butterworth, 2005; Mix, Sandhofer, & Baroody, 2005. See also Frank, Everett, Fedorenko, & Gibson, 2008.
[2]Frank et al., 2008.
[3]See Sarnecka & Gelman, 2004.
[4]Newcombe, Huttenlocher, & Learmonth, 1999.
[5]Mix, 2002.

[6]Huntley-Fenner, Carey, & Solimando, 2002; Miller, 1984.

[7]Noël, 2005; Shurtleff, Fay, Abbott, & Berninger, 1988.

[8]Ahlberg & Csocsán, 1999.

[9]Barnes, Smith-Chant, & Landry, 2005.

[10]Durkin, Shire, Riem, Crowther, & Rutter, 1986.

[11]Sarnecka & Gelman, 2004.

[12]See, e.g., Hurewitz, Papafragou, Gleitman, & Gelman, 2006; Kouider, Halberda, Wood, & Carey, 2006; Mix, Huttenlocher, & Levine, 2002b, pp. 136–139; Sarnecka & Gelman, 2004.

[13]See Lipton & Spelke, 2006.

[14]Skwarchuk & Anglin, 2002.

[15]Hecht, Torgesen, Wagner, & Rashotte, 2001; Stein, 2001.

[16]Fuchs et al., 2005. But see Jordan, Hanich, & Kaplan, 2003b.

[17]See Dowker, 2008; Gelman & Gallistel, 1978/1986.

[18]Fuson, 1988, p. 50; Mix et al., 2005.

[19]See Mix et al., 2005.

[20]Geary, Bow-Thomas, & Yao, 1992; LeFevre et al., 2006.

[21]Mix et al., 2002b, pp. 136–139.

[22]Fischer & Beckey, 1990.

[23]Fisher, Klingler, & Song, 2006; Mix et al., 2002b, pp. 29–32; Newcombe & Huttenlocher, 2000.

[24]Cheng & Newcombe, 2005; Uttal, Fisher, & Taylor, 2006.

[25]Newcombe & Huttenlocher, 2000.

[26]Skwarchuk & Anglin, 2002.

[27]For the following discussion, see, e.g., Cheng & Huntley-Fenner, 2000; Miller, Major, Shu, & Zhang, 2000; Miller, Smith, Zhu, & Zhang, 1995.

[28]Skwarchuk & Anglin, 2002.

[29]Zur & Gelman, 2004.

[30]Hudson, 1983.

[31]Bull, Marschark, & Blatto-Vallee, 2005; Nunes & Moreno, 1998.

[32]Shalev, Auerbach, Manor, & Gross-Tsur, 2000.

[33]Aram, Ekelman, & Nation, 1984; Cowan, Donlan, Newton, & Lloyd, 2005; Donlan, 1998; Hecht et al., 2001; Fazio, 1996; Manor, Shalev, Joseph, & Gross-Tsur, 2000.

[34]Aram et al., 1984; Donlan, 1998.

[35]Griffin, Case, & Siegler, 1994; Jordan, Huttenlocher, & Levine, 1992; Jordan, Kaplan, Ramineni, & Locuniak, 2008.

[36]Jordan et al., 2008.

[37]Stern & Stern, 1971.

[38]Fuson, Smith, & LoCicero, 1997.

Chapter 5

Reading and Writing Mathematics

Counting makes it possible to know values exactly and even to make some basic calculations. But counting alone does not get one very far; armed with only number words, one cannot calculate big numbers efficiently, remember the results easily, or formulate and communicate complex mathematical ideas clearly. For these reasons, literate society has developed a written symbolic system and teaches it to children in tandem with written natural language.[1] The notational system is efficient from a mathematical viewpoint, but learning and using it make significant perceptual and cognitive demands on the young student. This chapter reviews those requirements and the challenges they pose, as well as the system's most basic applications—arithmetic fact and algorithm mastery.

Reading and Writing Numerals

There are many ways to think about quantity. Children's earliest notions are in the form of concrete numerosities of objects or events (******). Mastery of the number-word sequence and the cardinality principle provide the phonological representation ("six"), and written alphanumeric notation adds symbolic (6) and orthographic (*six*) images. As all of this comes together around age 5 or 6, children think of the value abstractly as a place along the mental number line ('—'—'—'—'—'—*).[2]

These points may seem obvious, but they constitute the bedrock of conventional mathematics, just as written natural language rests on the alphabet. The specific format by which one takes in a number (e.g., Arabic or Roman numeral, written or spoken word) influences how efficiently one learns or uses it. For example, some people have trouble deciphering Roman numeral dates on cornerstones or the number-word dates and times on formal invitations. Studies have shown that people are slower to judge something as simple as whether a number is odd or even when it is written in English number words rather than in numerals, because they must transcode (i.e., mentally switch) it to a friendlier format first.[3]

Just as fluent reading depends upon the automatic association between the written letter *s* and the sound /s/ or between the written word *six* and the spoken word "six," the practical utility of various numerical expressions depends on how automatically one can link one numerical representation with another. For most mathematical purposes, the most critical association is between the symbol and its value (i.e., 6 ↔ '—'—'—'—'—'—*). Math relies just as heavily on the other associations, however. Learning to read and write numerals—making the 6 ↔ "six" connection—is a complex and often confusing task, just as is learning to read and write alphabetic text. After all, Arabic numerals, like the letters of the alphabet, are entirely abstract and arbitrary: Just as the shape of a letter provides no intrinsic clue about the sound it represents, so a numeral reveals no hint of the word or numerosity it symbolizes. (Unlike letters, however, numbers take on meaning by their position in the counting sequence, although most of the time one reads and uses them out of sequence.) Numerals, like number names, are conventions that students must learn; no amount of reasoning can lead one to the conclusion that 6 stands for (******) or even "six." By sheer force of memory, students must make the connections between the shape of the numeral, its place in the number sequence, the way its name sounds, and the numerosity it represents.

Numerical Decoding and Encoding

Being able to read and write numerals accurately and fluently—to "decode" and "encode" them, to draw a loose analogy to alphabetic reading—depends on several underlying perceptual and cognitive skills, in addition to the auditory processing abilities discussed in Chapter 4. For instance, reading numeric (and alphabetic) text fluently depends upon good visual skills. One of these skills is efficient visual scanning, which in turn depends upon the synchronized functioning of the muscles governing eye movement and focus. Poorly coordinated eye movements can take a significant toll on reading fluency and spelling.[4] Although the oscillating left-to-right eye movement necessary for reading is commonly assessed by asking individuals to rapidly scan and match digits, researchers have not yet investigated the relationship between mathematical competence and eye muscle function. More research is clearly needed in this area.

Children also require iconic visual skills to memorize the shape of each numeral and to recognize it as distinct among others that resemble it (e.g., 1 versus 7, 4 versus 9). To accomplish this, children need to remain alert to miniscule differences in lines and curves, while ignoring differences in font, size, color, or place on the page. They must also learn each numeral's correct left-to-right orientation, a task that requires both iconic and spatial skill. These are not easy tasks; children err more frequently on calculations involving digits that resemble other digits (e.g., 2 and 5, 6 and 9) than they do on those involving dissimilar-looking digits.[5] Children often confuse 3 with E for the same reason.

It is normal to generalize a perception to its mirror image across the vertical axis. After all, people would not be able to brush their hair in the mirror or recognize photographs of themselves otherwise. Initially, young children process letters and numerals in the same way as objects and pictures; hence, developing a reliable mental image of digits in their customary orientation involves learning to suppress that natural inclination. Unlike concrete objects, which have the same identity regardless of their orientation, abstract symbols derive their usefulness from their predictable orientation. Once a written form becomes a symbol and takes on meaning (e.g., numerosity, sound, relationship, operation, inflection), orientation matters. This is true for asymmetric letters, digits, and other symbols such as the inequality signs (> and <) and the question mark (?).

Orientation confusion affects how children write numerals, as well as read them. To write numerals correctly, students must not only recognize the correct form, but they must also be able to copy it and draw it from memory. Learning to write numerals correctly, like learning to recognize them, can take several years. In fact, many children go through a normal mirror-writing stage between ages 5 and 8 years.[6] If children do not have a firm grasp of reading and writing numerals, they will approach all written math cautiously and struggle to learn facts and algorithms. One study of 8- and 9-year-old students with severe math difficulties found that half of them wrote numerals backward, even without time pressure.[7]

A boy who flips digits + + + + + + + + + + + + + + + + + +

Douglas is a personable 15-year-old whose advanced vocabulary and mature conversation reveal exceptional intelligence. He was brought to our attention because he reverses numerals when he writes.

Though somewhat small at birth, his delivery, health, and early developmental milestones were unremarkable. He currently wears glasses to correct his visual acuity. Douglas was homeschooled until ninth grade. Throughout his school years, he had persistent reading, spelling, and writing difficulties. In second and third grade, he wrote his numbers, letters, and words "completely backwards," according to his mother. At age 9, his optometrist diagnosed ocular-motor impairments and gave him vision therapy for a year; his parents report that it "helped somewhat."

Despite this treatment, however, Douglas continued to reverse numbers when writing and had difficulty tracking lines of text. A psychological evaluation at age 14 revealed an exceptionally bright young man with superior verbal skills. By contrast, his nonverbal abilities were in only the average range. Most tellingly, his performance on tests of visual scanning measured below the 5th percentile for his grade, as did his spelling. In reading aloud, he confused *b* and *d* and rearranged letter sequences; his handwriting was poor.

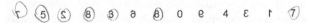

In math, Douglas still counted on his fingers and did not know what the division sign meant. A math evaluation at age 15 revealed mastery of the verbal counting sequence and basic awareness of quantity—he knew there are 7 days in a week and that $100 would be too much to spend on an ice cream cone, for example. He could subtract 3 from 35 in his head, as well as 3 from the serial differences, and could rapidly identify the larger of two numbers presented orally, suggesting good number sense.

Douglas had significant difficulty with written numerals, however. On a timed test requiring him to identify the pair of matching numerals in a row of similar-looking numerals, he performed accurately but very slowly (below the 15th percentile for his age). He also had difficulty rapidly identifying the correctly oriented digits in an array of correct and reversed digits.

After taking this test, Douglas seemed discouraged. He said that he found the test hard and that he often confuses the orientation of 5, 6, and 9 when he writes. His confusion was not limited to digits, however. He was just as slow on a test involving scanning nonsymbolic geometric forms, including marks resembling the math symbols + and ×, consistent with a general scanning impairment. In writing the alphabet under time pressure, he reversed both the *J* and the *K*. It is likely that Douglas's scanning impairments have hampered the establishment of correctly oriented symbol images in his brain.

How do these difficulties affect his math? Douglas has not fully mastered his multi-plication facts, even verbally. He also performed in the 3rd percentile for his age on a timed written test of mixed simple addition, subtraction, and multiplication facts, where he made only three errors but was very slow. His teacher confirms that his written math is laborious.

Douglas's mother is considering obtaining an optometric re-evaluation; renewing his ocular-motor exercises may yet improve Douglas's scanning. For symbol orientation, reading specialists use a host of (unproven) multisensory methods to help students dis-tinguish b from d, for instance, but there are no established techniques for training digit orientation. Faced with a student having difficulty with both symbol sets, one teacher we know just told him that if he could figure out the b, he could make his 6 just like it. And an experienced reading specialist, upon hearing our stories of digit reversal, ex-claimed that it can take 3 or 4 years just to disentangle b and d; tackling numerals as well would be extremely difficult. Clearly, more research is warranted.

Children must be familiar not only with what numerals look like, but also with their names (e.g., 6 ↔ "six")—for small and big numbers alike. When children first learn big numbers, they experiment with various ways to read them—365 as "thirty sixty-five," or 803 as "eighty-three," for example. Learning to read multidigit numer-als can take a long time, and young children at any grade level will show consider-able variability in this skill. Children's errors suggest that they often fail to connect the formal structure of the numerals to their base-10 significance, although their problems may also relate in part to the irregular form of the decade words (e.g., thirty) and to the silent role of 0. Of course, if children cannot read digits fluently, they will certainly struggle with multidigit numerals.[8]

Learning to instantly recognize a numeral in its correct orientation, to write it accurately and fluently, and to automatically associate that symbol with its proper name requires considerable practice during early education; those skills can remain fragile for quite awhile. Not surprisingly, in studies of primary-grade students, re-searchers found a significant connection between fluency in numeral recognition and naming on the one hand and math achievement on the other.[9] More research is needed on these associations in both young and older students.

Numerical Comprehension

To use conventional mathematical notation, one must know instantly what value a symbol represents (e.g., make the 6 ↔ '—'—'—'—'—'—* connection). To extend the reading analogy: How do students develop a fluent numerical reading vocabulary? To address this question, we revisit the relationship between symbolic number nota-tion and number sense.

By about age 5 years, children typically grasp the idea of ordered numerosities; their brains register them much as adult brains do. Children have learned the car-dinality principle of counting (i.e., that the last number counted tells "how many") and are starting to think of quantities in terms of their relationships to each other, as if ordered on a mental number line. But what about numerals—when are children able to read a numeral and automatically know its value?

A recent study addressed this question.[10] Researchers tested students at the be-ginning and end of first grade, as well as older students and adults. Participants took two timed tests in which they viewed pairs of unequal single-digit numerals. The numerals appeared in different-sized fonts so that the larger value looked physically bigger in some pairs (i.e., 3 4) and smaller in others (3 4). All participants were able to compare the values correctly. However, when the researchers asked them to compare the numerals' physical sizes—a nonnumerical, perceptual task for which the participants focused on the digits' appearance, not worth—only the beginning first graders could ignore the digits' values. For the older students and adults, the digits' values interfered with their ability to make this simple perceptual judgment (i.e., they said the 4 was physically bigger than the 3, even if it was physically smaller), suggesting that in their minds the values had become intrinsic to the symbols. The investigators concluded that first grade is a critical time for developing the association between symbols and their values, although children typically require a few more years to achieve complete numerical fluency. The researchers attributed this new link to increasing memory for the numerals and their relative values fostered by explicit teaching and practice in school; they concluded that it heralds the birth of a second sort of mental number representation founded on verbal associative memory.[11] This numeral knowledge enhances children's mathematical efficiency. For example, when they add 6 + 3 by the minimum addend procedure, they must not only compare 3 and 6 to know where to start, but they must also enter the mental number line at 6 without counting up to it, a step that requires certainty about the value of the numeral 6.

It is one thing to read single digits, but quite another to make sense of multidigit numerals. For example, when people compare two-digit numbers like 47 and 62, do they take into account the whole numeral (e.g., 47) or are the 4 and the 7 processed separately? In a timed experiment, students were asked to compare compatible (42 versus 57) and incompatible (46 versus 72) pairs—that is, 4 < 5 and 2 < 7 in the first pair, while 4 < 7 but 6 > 2 in the second pair. Based on how the children's response times varied, the researchers determined that the second graders processed the digits sequentially (the tens digit first, then the units), whereas the fifth graders did so in parallel (tens and units simultaneously)—a progression similar to that seen for letters and words in the development of fluent reading.[12] As with single digits, multidigit numerical comprehension may be partially independent of phonological input: Children with developmental language disorders who have adequate nonverbal skills can compare two-digit numbers, despite difficulty in naming digits. That is, the 6 ↔ '—'—'—'—'—'—'* association appears to be independent of the 6 ↔ "six" link, even for big numbers.[13]

Reading digits to the right of the decimal point can be especially confusing. When asked to locate decimal fractions on a 0–1 number line, the fifth and sixth graders in one study revealed two common misconceptions, both overgeneralizations from whole numbers: that more digits necessarily means a bigger value (i.e., 0.264 > 0.43) and that zeros to the left can be ignored (i.e., 0.07 = 0.7). The researchers found that either prompting the children to attend to the tenths digit or marking the number line in tenths helped them to see the numerals as place-value composites and thus understand their relative values.[14]

Arithmetic Facts

One of the most significant advantages of symbolic notation is that it facilitates remembering the results of many simple arithmetic transformations. In fact, the first math sentence most children learn to read is a simple arithmetic statement such as

2 + 3 = 5. Although young children initially learn arithmetic by counting objects and fingers, sooner or later they figure out that simply remembering the answer, if they can, is more efficient. Beginning with small sums in the early grades, the association between the addends and the sum may become automatic, such that mere exposure to a digit pair automatically brings their sum to mind.[15] In this way, children free up mental resources and arithmetic becomes a tool for problem solving rather than an end in itself.

How do children get from fingers to instant sums? Once-popular learning theories suggested that children progress neatly in a stepwise fashion through counting objects and fingers, counting up from the larger addend, decomposing complex sums into a series of simpler ones, and rote retrieval, until they solve all problems the mature way by instant recall. Recent studies, however, have put this development under a behavioral microscope and found that the path from counting to instant fact is slow and very uneven, with nearly as many steps backward as forward—a process that continues well into adulthood for many individuals.[16] Children who start out using their fingers to reckon sums and differences in kindergarten tend to be more accurate, however, and consistent accuracy leads to quicker mastery. Thus, among children who use their fingers early on, finger use declines by second grade as they master the facts. By contrast, children who do not use their fingers in kindergarten—a characteristic of many children from low-income homes—tend to rely on their fingers increasingly as they move through the primary grades and often fail to catch up to their more privileged peers in calculation accuracy. For reasons that are not yet clear, boys' finger use typically declines, and their accuracy improves, more rapidly than do girls'.[17]

In the course of learning to add or multiply, children not only memorize many facts, they also get better at using all of the strategies at their disposal. They also develop a reliable sense of which strategies will prove most efficient for which problems—judgments that shift as children become more competent and that often reflect individual differences in tolerance for inaccuracy versus perfectionism. Children who can muster a variety of strategies can deal with a wider range of problems and tend to learn better; in fact, some scholars have attributed Chinese children's arithmetic advantage to their more advanced strategy mix.[18] Moreover, studies have discovered what most people have known all along—namely, that in the fog of daily life, children and adults alike use whatever method works to get to the answer as quickly, accurately, and painlessly as possible. For adults more than children, the strategy of choice is rote retrieval, but alternatives often include finger counting, decomposition, guessing, and reaching for the calculator.[19]

Although problem solving enhances fact learning, mastery usually takes more systematic effort. Memorizing math facts can be daunting: There are 200 addition and multiplication facts involving transformations of only the numbers 0–9. (Children and adults tend to use derivative strategies for subtraction and division.[20]) Most math curricula require students to master these math facts by the time they are 10 years old. The next sections discuss how children commit these math facts to memory and which facts should be memorized.

How Children Remember

In learning math, students must remember three types of information: facts, procedures, and concepts. Facts consist of such things as basic arithmetic results and the value of π. Procedures are the steps in ordered activity, such as algorithms. Concepts include the reason one cannot divide by zero or why dividing by a fraction results in a bigger quantity. The brain represents these different facets of memory in different ways, but each supports the other.

Two aspects of memory play crucial roles in recalling math facts: association and the effectiveness of practice. Because human memory is associative, the brain stores associated information together; people remember information through its association with other things they know. As cognitive psychobiologist D.O. Hebb put it, "Neurons that fire together, wire together."[21] Thus memory is enhanced by association to familiar and predictable patterns. Rules, symmetry, repetition, rhythm, rhyme, and melody are all examples of regularities that help us remember information.

Sometimes pattern effects can be quite subtle. For example, although they are rarely taught this explicitly, children seem to pick up the parity (odd-even) patterns of arithmetic operations(e.g., in which the sum of an odd and an even number is always odd). One can see this effect in third graders learning multiplication, whose errors reflect addition parity patterns at the beginning of the year but who in time conform to the parity regularities of multiplication.[22] Unfortunately, associations can also impede fact mastery. For example, numbers that appear near each other on a fact table (i.e., $6 \times 5 = 30$; $7 \times 5 = 35$) in the same position on another table ($6 + 5 = 11$; $6 \times 5 = 30$) or are otherwise visually associated (4×5, 45, 67) can lead to confusion. In fact, most fact-retrieval errors tend to stem from table confusion, particularly among third- and fourth-grade students as they move from addition to multiplication.[23]

"Agnes" comic strip reprinted by permission of Tony Cochran and Creators Syndicate, Inc.

Brain studies have shown that rehearsal (i.e., use, practice, and drill) increases the long-term strength and efficiency of circuits managing the storage and retrieval of information in long-term memory.[24] The effectiveness of practice has several implications. First, when elementary school workbooks overrepresent easy facts and algorithms at the expense of hard ones, as some do, children may never master the hard ones.[25] Second, if students drill without corrective feedback, they are likely to store errors in long-term memory. Third, the long-range value of focused practice is not limited to arithmetic: In a lifespan study of basic high school algebra and geometry retention, researchers found that only those students who subsequently continued their math education and used their basic skills in college retained these fundamentals, whereas students who aborted their algebra and geometry study after high school experienced a pronounced and steady decline in those skills from that point on—even if they later used math on the job. These findings led the researchers to conclude that the key to lifelong math retention is a prolonged period of purposeful practice, use, and relearning in the academic setting. For this reason, students tend to remember arithmetic better than the later-learned advanced skills, because they learn it over the protracted time period of formal elementary and secondary education.[26]

Arithmetic information has three familiar formats—phonological (i.e., "three plus two equals five"), symbolic ($3 + 2 = 5$), and conceptual ('——'——'——*——'——*)—providing three different ways to remember it: verbally, iconically, and spatially, respectively. For example, children who learn their facts by reciting tables, like those who learn the alphabet by verbal repetition, can locate a fact from a cluster of phonologically associated number words: Hearing "three," "two," and "five" may trigger recall of the four addition and subtraction facts relating these three numbers. Because verbal information is necessarily sequential, however, misremembering a fact may require the child to return to the start of the sequence, thereby demanding additional working memory and time. Moreover, relying on verbal skill and memory to learn and recall arithmetic facts leaves children who have language disorders especially vulnerable. Children can also learn the way the notational statements appear on the page (i.e., $3 + 2 = 5$), with the numerals' visual image serving as the mnemonic trigger. If children learn facts orally and are tested in writing, they must mentally translate from one format to the other prior to responding, as they would with the reverse approach.

Because memory is associative, meaningful information is remembered better than arbitrary facts; thus the conceptual recall mode is perhaps the most important. Meaningfulness necessarily relates to personal experience: A chess expert, for example, easily recalls a chess play that a nonplayer, who finds it random, cannot remember.[27] Not understanding the game makes the facts hollow and hard to remember for the nonplayer. Studies have shown that children remember arithmetic facts more accurately and reliably and can apply them more flexibly if they have learned them by way of strategies—for example, $6 + 7 = 6 + 6 + 1$ or $9 + 5 = 10 + 4$, both accomplished by decomposing and recomposing the addends to simplify the problems. This approach keeps students' attention focused on the values and their relationships to one another, not just the numbers' phonological or iconic characteristics. In fact, strategic reasoning applied on the job may explain the pockets of extraordinary math expertise acquired by people who are either unschooled or unable to learn the same skills by rote in the classroom.

Factual, procedural, and conceptual knowledge and memory support each other.[28] Thus, using a mix of learning and drill strategies should maximize the chances that students master this plethora of information. Having access to alternative approaches is particularly crucial for students with impairments in any of the necessary underlying skills.

Which Facts?

Do students really need to memorize all of these math facts? Because addition is commutative (i.e., $4 + 3 = 3 + 4$), the table has considerable redundancy. Although the same can be said for multiplication, it is important to note that in multiplication, the first number functions as a multiplier, whereas the second represents the amount that is to be multiplied. Therefore, 4×3 does not mean the same as 3×4 because four groups of three things is not the same as three groups of four things—

a conceptual distinction that becomes salient in algebra. What the two statements have in common, however, is the same resulting amount; the knowledge that the two problems can be collapsed into one in this sense may help a student who is struggling to remember what that result is.

Table redundancy raises the possibility that students may more efficiently focus on memorizing only half the facts in each table and derive the others using the commutative property. But which half of the table should they learn? One clever study found that, although they are not aware of doing it, adults retrieved addition facts by first comparing the two addends, just as they had when using the minimum addend strategy as children; regardless of the way the individuals had been taught the table, the answers were filed in memory with the larger addend coming first. (For the "zeroes" and "doubles," participants went directly to the rule or fact governing those sums.)[29]

Addition, like subtraction, is easily visualized on the mental number line, which may be why the strategy used by most people is so close to finger-counting methods. Multiplication is a bit more complex. Young children find it easier to skip count by the larger operand (e.g., $6 \times 3 =$ "six, twelve, eighteen") when the other one is very small, but as the values increase and as children gain experience, operand order has less effect on efficiency.[30] Schools in several countries, including China, teach only the smaller-operand-first half of each table; notably in one study students were slower on problems presented with the larger operand first because they needed to mentally transpose the digits before solving. However, they were nearly as competent on the large-number facts as on the small-number ones, suggesting that they had learned more diverse facts.[31]

In all, these preliminary findings open several important questions. First, does fact mastery depend on the way lessons organize the material? Second, does one's ability to retrieve arithmetic facts depend on the method by which one learns them or on the format in which problems are presented? Third, is it possible to devise an optimal individual learning and drill strategy based on a student's particular problem-solving style and perceptual and cognitive strengths? The answers to these questions await further research.

Symbols and Algorithms

Conventional mathematics, like written natural language, is a formal system of symbols, syntactic rules, and procedures that provide an efficient and consistent way to formulate, remember, express, and communicate ideas. Some of these expressions, such as the simple arithmetic statement (i.e., *3 + 4 = 7*) and question (*3 + 4 = ?*), are a natural outgrowth of basic number knowledge and set the groundwork for complex arithmetic and algebraic algorithms. Other conventions, however, are unintuitive extensions of those rules—for example, fractional exponents and negative multipliers.[32]

The plethora of symbols, expressions, and algorithms may seem overwhelming to some students. For example, most punctuation marks serve different functions in alphabetic and numeric text.[33] Moreover, by the time students reach high school, they have learned at least 14 ways to express the four arithmetic functions. As students move beyond whole-number arithmetic, symbols proliferate; in fact, some advanced college math textbooks have very little natural-language text at all. This section examines two of the most common frustrations introduced by written mathematical conventions: the equals-sign problem and "buggy algorithms."

One of the symbols that children find most vexing is the lowly equals sign ($=$), which can also be rendered as the "bottom line" under the operands in vertical arithmetic operations or even as an underscore when included in other relational symbols (e.g., $\geq$, $\leq$). Studies have found that many elementary and middle school students think of $=$ as simply another operational sign—in this case meaning *now do the problem*— or as a word, like *total* or *altogether,* that signals the end of a word problem. Translating $=$ as the action word *makes* (e.g., *two plus one makes three*) or *leaves* (*two minus one leaves one*) further suggests an operational, rather than relational, interpretation. Thus, $3 + 2 = 2 + 3$ makes no sense to some students because the plus sign ($+$) in the "answer" persuades them that they have not finished doing the problem.[34] For this reason, some educators advocate translating $=$ as *is the same (number, amount) as* or using a different symbol (e.g., $\rightarrow$) to stress the operation.[35] Unfortunately, even middle school textbooks rarely use $=$ in any form but the standard *operation = answer.*[36] This limited understanding of $=$ leaves students unprepared for algebra.[37]

Another common stumbling block is procedural: Children often misinterpret, only partially understand, or systematically overgeneralize arithmetic algorithms and fail to connect them to what they know from experience about numbers. For example, by following the rule "Subtract the smaller number from the larger number," children arrive at $13 - 6 = 13$ by diligently subtracting 3 from 6. These systematic errors, or bugs, lead to what researchers refer to as "buggy algorithms"—that is, the results of well-intentioned children trying to do what they thought they had been told or inventing clever shortcuts on long assignments.[38] Multidigit subtraction is particularly susceptible, perhaps because one "borrows" from left to right while operating from right to left.

What were they thinking? $+ + + + + + + + + + + + + + +$

The following "buggy algorithms" are real examples culled from our students. They were from a page of subtraction; the student gave up on the last example before it was finished:

$$\begin{array}{r} \$40.00 \\ -\ 29.25 \\ \hline \$29.25 \end{array} \qquad \begin{array}{r} 6\overset{1}{2}.4 \\ -\ 8.207 \\ \hline 160.207 \end{array}$$

$$\begin{array}{r} \overset{3\,4}{\cancel{5}}00 \\ -206 \\ \hline 104 \end{array} \qquad \begin{array}{r} \overset{6}{\cancel{8}}\,1 \\ 2\overset{}{8}5 \\ -\ 187 \\ \hline ?\ \ 8 \end{array}$$

In this final example, the student became so befuddled, we show her steps sequentially. When her work did not check, she gave up.

$$\begin{array}{r} \overset{8}{2}9\overset{1}{3} \\ -\ 197 \\ \hline 6 \end{array} \ \longrightarrow\ \begin{array}{r} 2\overset{7}{\cancel{9}}\overset{1}{3} \\ -\ 197 \\ \hline 6 \end{array} \ \longrightarrow\ \begin{array}{r} \overset{0\ \ 7}{\cancel{2}\cancel{9}}\overset{1}{3} \\ -\ 197 \\ \hline 86 \end{array} \qquad \begin{array}{r} Check. \\ 197 \\ +\ 86 \\ \hline 283 \end{array}$$

A special case of buggy algorithm can be found in the Algebra I classroom. Many students new to algebra frequently produce overgeneralizations of basic correct rules: $(x + y)^2 = x^2 + y^2$ as an erroneous extension of $(xy)^2 = x^2y^2$, for example. In examining these errors more closely, however, researchers discovered that they are based on aesthetic, rather than conceptual, considerations, satisfying a need for a kind of dynamic visual resolution. As the researchers note, these errors do not reflect a misconception but rather a lack of any conception. They also found that boys had relatively little trouble avoiding this trap, whereas girls accounted for most of the errors.[39] Future research should investigate whether this kind of early algebra difficulty is related to an iconic visualization style focused on the symbols themselves, rather than on their underlying meanings.

Reading Disability and Mathematics

With so many similarities between reading alphabetic text and deciphering mathematical notation, one might expect that children who excel (or struggle) in one area would also excel (or struggle) in the other. Indeed, the behavioral genetics studies discussed in Chapter 1 also found a very large genetic overlap between math and reading, suggesting that the underlying cognitive abilities contributing to both are probably governed by the same set of genes. The continuity between ability and disability makes it likely that impairments in the two domains are also genetically linked. The correspondence is not perfect, however; in fact, about one third of the genetic influence on math skills is not related to reading or even to general intelligence, although the affected math-specific skills have not yet been determined.[40]

Given the partial nature of the genetic relationship, it is not surprising that research on the relationship between disabilities in reading and conventional math skills has produced mixed results, with important questions as yet unresolved. As expected, studies have found a high co-occurrence of dyslexia and math disabilities. Researchers estimate the math disability rate among children with specific reading disabilities is about 35%; adults with reading disabilities have an estimated concurrent math disability rate of more than 70%. Conversely, among children with math disability, researchers estimate that 17%–66% have a concurrent reading disability; comparable estimates for adults run 39%–48%.[41] (The broad ranges stem from the studies' different working definitions and data sources.) Although the disability overlap is high, these studies also support the genetic finding that a substantial proportion of children with significant math difficulties do not have reading disabilities. Thus, identification of students meriting special academic services for math cannot be based on reading tests. Conversely, students with reading disabilities should be routinely screened for math difficulties and vice versa.

Although students with dyslexia tend to perform worse on reading and spelling than on math, their math skills stall out earlier than reading in school. Several studies found that among individuals with dyslexia, grade-normed math scores tumbled from near the 40th percentile in the intermediate and middle school years to below the 10th percentile by early adulthood, when they still measured at the fourth-grade level, although some scholars dispute that persistence of severe math disability is related to reading.[42] One study of fifth-grade students with dyslexia demonstrated that well-established reading instruction techniques produced significant reading improvement but only modest gains in calculation fluency.[43]

Several large-scale investigations have revealed that many students with dyslexia perform especially poorly in math even when reading the problems is not an issue. Some students with dyslexia do, of course, have trouble reading math problems and extracting relevant information from them. Many lack adequate vocabulary or syntax or become confused by notation. But most often, these students do not acquire knowledge or fluent use of arithmetic facts as young children, thereby setting the foundation for math difficulties later. Many children with dyslexia have particular difficulty recalling multiplication facts, perhaps because multiplication facts (more than the others) are learned by rote and draw on phonological functions, numeral decoding, and verbal memory.[44] There is little scientific consensus about the source of this poor fact mastery; some researchers point to underlying phonological impairments and others to faulty visual processing, but consistent evidence is lacking.[45]

Do difficulties in reading letters and words imply difficulties in reading numerals? Research has shown that students with dyslexia frequently have trouble with written calculations, even though the numbers are in numerical (e.g., *6*) rather than orthographic (*six*) format; many of these students have a poorly established sense of what numerals look like. One recent study found that nearly one in three students ages 8–10 years who were referred for a general learning-disability evaluation had significant difficulty recognizing the correct orientation of numerals; they had the same confusion about letters and were particularly poor readers.[46] Even more commonly, individuals with dyslexia often cannot process numerals fluently (i.e., quickly and automatically), not only in recognizing their correct orientation but also in naming them.[47]

One theory of dyslexia is that it represents a disruption of "fast-cell," automatic brain processing.[48] A study demonstrated that children with dyslexia took four times longer than did those without the disorder to remember an array of single digits; another study showed they were abnormally slow at recognizing the correct orientation of digits.[49] In the course of learning to read, most typically developing children go from processing letter strings sequentially to the more efficient method of processing them simultaneously (*c . . . a . . . t → cat*). Multidigit numeral reading follows the same development (*2 . . . 8 → 28*).[50] Individuals with dyslexia are unable to process words efficiently, but it is unclear how they process multidigit numerals. The possible effect of dyslexia on learning nonnumerical mathematical notation, such as operational and relational signs, also has not been determined.

It is not unreasonable to wonder whether disrupted automatic processing affecting all symbols may explain both reading and math disorders. Research, however, suggests that this is not necessarily the case: Not all students with reading disorders struggle with math symbols and some students without reading disorders do. Some children with math difficulties (with and without dyslexia) name numerals slowly and/or reverse them in writing, regardless of reading ability. By contrast, some students with dyslexia but no math difficulties do not reverse digits or have any difficulty naming them quickly.[51] Other students, with and without reading disabilities, fail to grasp the quantitative significance of numerals—that is, to make the $6 \leftrightarrow$ '—'—'—'—'—'—'* connection. Although these children have no trouble identifying the larger of two sets of dots, say, they are baffled when asked to compare 5 and 4.[52] Evidence also suggests that some students whose dyslexia impedes fact mastery can improve their multiplication fluency over time with practice and knowledge of the rules and regularities governing it, indicating they may process multiplication independently of the natural language reading circuit or can marshal independent compensatory strategies.[53] Thus the link, if there is one, between reading and math disorders is not straightforward. Chapter 6 revisits this issue in examining the relevant brain functions.

Regardless of the relationship between reading and math disorders, however, these findings highlight the importance and difficulty of learning basic mathematical reading and writing. Researchers must ensure that they do not take these skills for granted and attribute students' failures on notational math exercises to other factors, such as weak number sense or poor memory, without first ruling out problems in numeral decoding and encoding. If children do not have a firm grasp of the notation, they will approach all written math cautiously and find learning quantitative relationships, numerical patterns, arithmetic facts, and algorithms frustrating. The challenge is not unlike learning to read and write a second language in a different alphabet. Just as students with dyslexia read slowly because they struggle to recognize letters and words and match them to their sounds, so many pupils with math difficulties perform poorly on math tasks that require instantly recognizing, naming, and knowing the values of written numerals. Students with hand-eye coordination impairment also struggle to organize the notation on the page, making it difficult for them to show their work.[54] For all these reasons, students with math difficulties, like those with dyslexia, often experiment with inefficient strategies. The difficulty of mastering fundamental mathematical reading and writing should not be underestimated.

Classroom Implications

Children's comprehension of symbolic arithmetic transformations depends on their knowledge of the signs and symbols of mathematical language. Without a secure grasp of + and =, as well as the numerals *6, 2,* and *8* (including their orientation and order on the page; their names; what value, operation, or relationship they stand for; and how the numerals' values relate to each other), no child can make sense of *6 + 2 = 8.*

Students have two goals as they learn to read, write, and calculate with numerical notation: accuracy and fluency. These aims parallel those in learning natural-language reading and writing. Unlike the field of reading and spelling instruction, however, mathematical language does not yet have established pedagogical methods. It is not yet known what teaching methods work most effectively for whom. Nevertheless, some educators have taken a page from the reading experts and generated class-room approaches that seem promising.

To teach reading and writing to children with either general language disorders or a specific reading disability, reading experts advocate multisensory techniques con-sisting of visual, auditory, tactile, kinesthetic, and/or speech components.[55] For chil-dren with comparable math difficulties, these methods might include tracing numerals over dotted outlines, using rhythm or different colors for the pencil strokes, "writing" the numerals with the fingertips on a rough surface, drawing the numerals with a stick in sand, fitting plastic numeral forms into fixed cutouts, and associating iconic or narrative mnemonics with each digit. Other techniques include explicit instruction in arithmetic syntax and using graph paper to keep algorithm components aligned. To remember procedures, some students rely on entertaining mnemonics, such as *Please Excuse My Dear Aunt Sally* (order of operations) and *Dead Mice Smell BaD* (long division), or attaching operations to story lines.[56] None of these methods have been rigorously tested, however, and there are many potential pitfalls. For students with language or reading disorders who have phonological sequencing difficulties, verbal mnemonics may prove as difficult as memorizing the order of operations, for example, whereas iconic imagery may distract students and inhibit generalization.

As to the visual component, optometrists suggest that being able to discrimi-nate mirror-image symbols depends on a robust awareness of left versus right on stu-dents' own bodies, which they believe can and should be taught as a prereading skill.[57] The field of reading and dyslexia has long recognized the necessity of system-atically coordinating the underlying perceptual and motor skills, as well as establish-ing crucial associations—a need sometimes overlooked in mathematics education. Scientific research on the effectiveness of these untested teaching methods is clearly warranted.

To firmly affix numerals to their values (and names) and to demonstrate how they compare to each other, some educators prefer to teach reading and writing nu-merals in tandem with counting-board, number frame, or number-line illustrations of their values. By this method, students learn simultaneously to read the numerals and locate them correctly (Figure 5.1; also see Figures 4.1 and 4.2). Then they learn to write the numerals, as well as mathematical signs, and use them to describe the values and their transformations. In this manner, children learn to associate the numeral *6,* for example, with its place on the number line; six unit cubes, the 6 block, and six grooves in the sixth column on a counting board; and the result of small-digit additions and subtractions in a number frame. Those who advocate using spatial number analogs believe that these rich associations—built up from a wealth of their earlier pre-reading counting board activity—implant firmly in children a solid understanding of numbers. They learn to write numerals to describe what they see; facts and algorithms emerge as a natural record of the physical manipulations.[58]

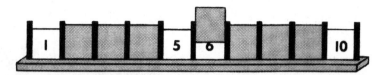

Figure 5.1. Learning to associate numerals with ordered values on a number line. (From CHILDREN DISCOVER ARITHMETIC: AN INTRODUCTION TO STRUCTURAL ARITHMETIC, 2ND ED. by CATHERINE STERN and MARGARET B. STERN. Copyright © 1949, 1971 by Harper & Row, Publishers, Inc. Reprinted by permission of HarperCollins Publishers.)

Base-10 concepts are commonly introduced with dimensional extensions of number blocks—100-flats and 1,000-cubes. These are not linear analogs of numbers but do serve to illustrate place-value concepts and multidigit numeral reading and writing. One study used a sequence of number-line instruction followed by base-10 instruction to teach multidigit addition and subtraction to second graders. The researchers found that the number line fostered sequential strategies (e.g., $45 + 39 = 45 + 30 + 5 + 4 = 84$), while base-10 instruction enhanced decomposition approaches ($45 + 39 = [40 + 30] + [5 + 9] = 70 + 14 = 80 + 4 = 84$). They also found that stressing these arithmetic concepts along with algorithmic competence led to a much more flexible approach to problem solving than teaching the algorithms in isolation. In this manner, students gained the tools to choose their strategies based on problem characteristics (e.g., base-10 procedures work best for addition, while a number line approach helps avoid errors on complex subtraction).[59] Preliminary trials of new software designed to improve number sense significantly increased subtraction accuracy of young children with math disabilities and may be a promising pedagogical alternative that merits further research.[60] The relative advantages and sequencing of number- line and base-10–block instruction for each of the arithmetic operations may also be worth investigating.

Arithmetic competence requires fluency, as well as accuracy. To achieve fluency, children need to practice the relevant facts until they can retrieve them automatically. Research has demonstrated that distributed (spaced) practice is a more potent method of retaining both facts and mathematical procedures than massed practice (cramming). Thus children should continue to practice previously learned facts as they gradually learn new ones. Researchers have also discovered that retrieval practice (as in quizzing) along with corrective feedback is more effective than further studying. Some drill software has proven effective for elementary school students of all ability levels.[61]

Drill should not be divorced from procedural and conceptual instruction, however. To the contrary, drill combined with strategy instruction is essential and ensures that students can apply their factual knowledge flexibly to more complex calculations, word problems, and estimation. Many educators recommend that students begin with the most easily learned facts that obey rules or that generate memorable patterns, such as doubles or squares, and then derive the "hard" facts from those. Leaving the hard facts for last does not mean that teachers and textbooks should give them short shrift in subsequent drill and calculations, however. In fact, some educators recommend presenting them even more often than the easy ones. Discussions of rules and strategies should be accompanied by systematic practice and review.[62]

A pedagogical question often asked by frustrated teachers, parents, and students is, "Why not just let children use calculators?" There are good reasons for doing so: Because obtaining an answer by calculator is quick, a student can solve more problems and thus get more practice; also, the answers are more likely to be correct, thereby boosting familiarity and fact mastery. The research indicates that in the short term, calculator use has no significant effect on problem solving or conceptual development. One study of third graders learning multiplication has shown, however, that manual computation is superior for mastering procedures, although calculators may benefit those who have already done so, for the reasons just indicated.[63] Studies on long-term effects are sparse, but suggest that calculator use in elementary and secondary school may have negative consequences in college, perhaps because students lack the extended procedural practice. One calculus professor has raised the concern that "students may not get the necessary feel for the number system needed to progress in math."[64] More research on the nature and extent of calculator use in the lower grades and its long-term effects is warranted.

Another open question is the degree to which the student or the teacher should direct the learning process. The results of one small study of typical French second graders new to multiplication reveal that students use the strategies that work for them, regardless of their teacher's pedagogical emphasis or role in directing instruction. In this case, the classroom approach was rote drill, but the students actually used the common methods of repeated addition and guessing when rote memory failed. The researchers concluded that "many characteristics of arithmetic learning reflect the nature of the system doing the learning rather than the instruction per se."[65]

The most effective instruction will adapt to the "system doing the learning," particularly with struggling students. Researchers recently examined the role of pedagogy in teaching multiplication facts to two groups of low-achieving 8- to 11-year-old students. For one group of students, the teacher structured the class discussion of particular multiplication facts, but only strategies "discovered" by the children were admissible; the teacher never demonstrated any strategy. For the other group, the teacher explained and demonstrated a small set of strategies and told the children how and when to apply them; although students occasionally arrived at alternative approaches, they were strongly encouraged to use the teacher's methods. The investigators found that the two groups became equally fluent and accurate over the study period (more than classmates not taught by either approach) and used a comparable range of strategies. However, the group receiving explicit instruction better applied their skills to bigger numbers and word problems. The researchers concluded that struggling students benefit from some explicit instruction.[66]

A final question for future research involves the format in which children should drill arithmetic facts. Educational traditions vary, with some preferring oral drill and others flash cards, computer, or written drill— a variable that researchers do not always take into account. For some children, format may not make a significant difference. For students with perceptual or cognitive impairment, however, format may make an enormous difference. Some investigators have recommended that children with phonological or other severe language impairment, for example, drill math facts using visual computer games.[67] Other learning disability specialists advocate drilling the facts in both phonological and symbolic modes, by reading, writing, and reciting each fact, or even setting facts to rhythm, rhyme, or song. Multisensory instruction is a common feature of the most successful remedial reading programs, but the efficacy of these methods for math or reading remains to be demonstrated scientifically.[68]

Conclusion

Mathematical language has a symbol system, vocabulary, and grammar that is mathematically elegant, but complex and cognitively demanding to learn. Some of the skills necessary to master it are the same as those required to learn alphabetic reading, whereas others are unique to mathematics. Like natural-language reading and writing, mathematical language draws on visual and phonological faculties and requires the ability to integrate them with fine-motor skills. Although the brain registers quantity the same way regardless of format, students must read and learn the notation in several different forms. For many students with dyslexia, mathematical language poses additional challenges. The overlap between mathematical and alphabetic reading is incomplete, however, and many students who read well nevertheless struggle with mathematical reading and writing.

In considering the direction of future research, it will not be enough to determine the appropriate pedagogical approaches for each math skill. The methods that work best for any given set of cognitive and perceptual strengths and weaknesses require rigorous investigation. There is clearly plenty of need and opportunity for future research.

Notes

1 Munn, 1998.
2 Dehaene, 1992.
3 Dehaene, Bossini, & Giraux, 1993.
4 Kulp & Schmidt, 1996; Stein, 2001.
5 See Heathcote, 1994, as cited in Bull, Johnston, & Roy, 1999.
6 Dehaene, Cohen, Sigman, & Vinckier, 2005; Lachmann, 2002.
7 Landerl, Bevan, & Butterworth, 2004.
8 Power & DalMartello, 1997; Skwarchuk & Anglin, 2002.
9 Boone, 1986; Geary, Hoard, Byrd-Craven, Nugent, & Numtee, 2007; Murphy, Mazzocco, Hanich, & Early, 2007.
10 Rubinsten, Henik, Berger, & Shahar-Shalev, 2002.
11 Rubinsten et al., 2002. See also Girelli, Lucangeli, & Butterworth, 2000; van Galen & Reitsma, 2008.
12 Nuerk, Kaufmann, Zoppoth, & Willmes, 2004.
13 Donlan & Gourlay, 1999.
14 Rittle-Johnson, Siegler, & Alibali, 2001.
15 LeFevre, Bisanz, & Mrkonjic, 1988; Lemaire, Barrett, Fayol, & Abdi, 1994.
16 See Siegler, 2003.
17 Jordan, Kaplan, Ramineni, & Locuniak, 2008.
18 Geary, Fan, & Bow-Thomas, 1992.
19 LeFevre et al., 1996; Siegler, 2003.
20 Campbell & Xue, 2001; Robinson et al., 2006.
21 D.O. Hebb, as cited in Casey, Giedd, & Thomas, 2000, p. 246.
22 Lemaire & Siegler, 1995.
23 Lemaire et al., 1994; Miller & Paredes, 1990.
24 Casey et al., 2000.
25 Geary, 1996.
26 Bahrick & Hall, 1991.
27 Chase & Simon, 1973.

[28]Delazer et al., 2005.

[29]Butterworth, Zorzi, Girelli, & Jonckheere, 2001.

[30]Butterworth, Marchesini, & Girelli, 2003.

[31]LeFevre & Liu, 1997.

[32]Pimm, 1987, p. 186.

[33]Pimm, 1987, p. 141.

[34]Pimm, 1987, pp. 187–188. McNeil & Alibali, 2005; Seo & Ginsburg, 2003.

[35]Stern & Stern, 1971.

[36]McNeil et al., 2006.

[37]Knuth, Stephans, McNeil, & Alibali, 2006.

[38]Hatano, Amaiwa, & Inagaki, 1996.

[39]Kirshner & Awtry, 2004.

[40]Plomin, Kovas, & Haworth, 2007.

[41]For child rates, see Badian, 1983; Barbaresi, Katusic, Colligan, Weaver, & Jacobson, 2005; Lewis, Hitch, & Walker, 1994; Mazzocco & Myers, 2003; Gross-Tsur, Manor, & Shalev, 1996. For adult rates, see McCue, Goldstein, Shelly, & Katz, 1986; Shafrir & Siegel, 1994.

[42]Frauenheim, 1978; Shalev, Manor, & Gross-Tsur, 2005; Trites & Fiedorowicz, 1976.

[43]Meyler, Keller, Cherkassky, Gabrieli, & Just, 2008.

[44]Miles, Haslum, & Wheeler, 2001; Simmons & Singleton, 2006; Turner-Ellis, Miles, & Wheeler, 1996.

[45]Hecht, Torgesen, Wagner, & Rashotte, 2001; Jordan, Hanich, & Kaplan, 2003a; Simmons & Singleton, 2008; Turner-Ellis et al., 1996.

[46]Badian, 2005.

[47]Terepocki, Kruk, & Willows, 2002; Wolf, Bowers, & Biddle, 2000.

[48]Stein, 2001.

[49]Terepocki et al., 2002; Ellis & Miles, 1977.

[50]Nuerk et al., 2004.

[51]See, e.g., Landerl et al., 2004.

[52]Iuculano, Tang, Hall, & Butterworth, 2008; Rousselle & Noël, 2007; Rubinsten & Henik, 2005.

[53]Turner-Ellis et al., 1996.

[54]See, e.g., Rourke, 1993.

[55]Moats & Farrell, 2005.

[56]Manalo, Bunnell, & Stillman, 2000.

[57]McMonnies, 1992.

[58]Stern & Stern, 1971.

[59]Blöte, Van der Burg, & Klein, 2001.

[60]Wilson, Revkin, Cohen, Cohen, & Dehaene, 2006.

[61]Gersten et al., 2008; Hasselbring, Goin, & Bransford, 1988; Mayfield & Chase, 2002; Pashler, Rohrer, Cepeda, & Carpenter, 2007.

[62]Tournaki, 2003; Woodward, 2006.

[63]Rittle-Johnson & Kmicikewycz, 2008.

[64]Wilson & Naiman, 2004, p. 119. See also Campbell & Xue, 2001; Gersten et al., 2008.

[65]Lemaire & Siegler, 1995, p. 95.

[66]Kroesbergen, Van Luit, & Maas, 2004.

[67]Fazio, 1996; Hasselbring et al., 1988.

[68]Moats & Farrell, 2005.

Chapter 6

The Brain and Conventional Mathematics

H ow does the brain "read" mathematical notation? How does it know that the numeral *6* is meaningful and not just a squiggly line, and that it is associated with the spoken word "six" and represents a certain quantity? How does the brain read multidigit numerals and arithmetic statements? How does it estimate and calculate? Does it learn subtraction the same way it learns multiplication, and how does it remember arithmetic facts? Mathematical and natural language have much in common: Does the brain process these two languages and symbol systems the same way? If not, where do they diverge? This chapter addresses these questions, among others.

Our knowledge of the brain's conventional number skills comes from various sources, including experimental, neuropsychological, and functional brain imaging studies of individuals with adequate math skills and of those with math disabilities stemming from injuries and "split-brain" conditions. Additionally, experimental and brain imaging studies of children have elucidated maturational changes. It turns out that the mathematical language brain circuits are extraordinarily complex, varying even across natural languages and cultures,[1] and the scientific questions are far from settled. The mind–brain relationship is a hot research topic; many new discoveries are expected in the coming years.

In navigating around the brain, we use two reference points: the IPS, which govern our quantitative understanding, and the natural-language and reading circuit. Because conventional mathematics ultimately deals with numerical value, the IPS are key. Few systematic neuroscientific studies have tried to link math and reading directly; this chapter examines the available work to determine where they overlap or intersect.

Although both math and reading involve significant activity in the front portion of the brain, which controls important general functions not specific to these skills, this chapter focuses on the back regions. (Section III will explore the role of the frontal areas.) These back regions consist of the temporal, parietal, and occipital lobes (Figure 6.1). Generally speaking, it is the job of the surface, or cortex, of these lobes to receive, organize, and integrate incoming information gathered through all the senses. Incoming math information includes spoken words; written words and symbols; and the sight, sound, and feel of things to be measured or counted, including fingers. The back regions of the brain also play a significant role in allocating visual attention—a critical component both of counting and of directing attention along the mental number line, which is itself a parietal function. These various functions are neuronally linked with one another.[2]

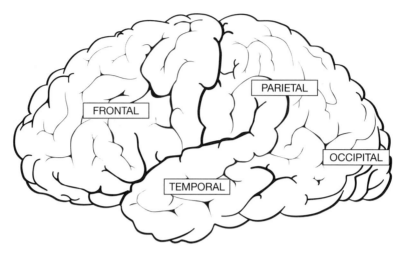

Figure 6.1. The four major lobes of the brain, shown here in the left hemisphere.

Furthermore, the brain divides down the middle, front to back, into two very similar hemispheres that communicate with each other via nerves that connect them (Figure 6.2). Natural language—including comprehension, speech, reading, and writing—is largely confined to the left hemisphere. (Speech is technically a frontal lobe function, but it is included in this discussion of language.) "Split brain" and brain injury studies have shown that the ability to read and write numerals (6 ↔ "six") and count—the essential building blocks of formal mathematics—is also under the control of the left hemisphere, suggesting a fundamental link between the two languages.[3] (Some individuals have language in both hemispheres, but the relation between atypical hemispheric organization and math ability is complex and somewhat speculative and thus beyond the scope of this brief chapter.[4]) The left parietal lobe in particular plays an important role in calculation and perhaps arithmetic fact storage. This chapter explores the mature brain's representation of numerals, its activation during arithmetic, the development of math functions in the brain, and the educational implications of the research findings to date.

How the Brain "Reads" Numerals

Chapter 5 discussed the foundational skills required for notational arithmetic: recognizing, naming, and determining the value of written numerals. These skills are often taken for granted because they come so easily to many people. However, as one can discern from Figure 6.2, the brain circuitry underlying these skills is quite complex.

Visual Recognition

When a person sees a numeral (6), the stimulus is sensed by the occipital cortex in the very back of each hemisphere. From there, the information is transmitted along the iconic visual pathway to an area in the left hemisphere, where the brain recognizes it as a numeral, distinct from meaningless squiggles. This small cortical region, known as the visual word-form area (VWFA), got its name because it also recognizes

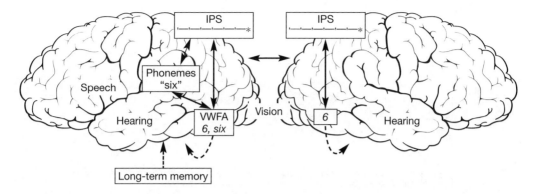

Figure 6.2. Mathematical language functions in the brain. (*Key:* IPS = intraparietal sulcus. VWFA = visual word-form area.) (*Source:* Dehaene, 1997.)

letters and familiar spelling patterns. Other parts of the brain can analyze letters, words, and numerals, but only the VWFA can learn to do so instantaneously, as required for efficient reading and calculating. The capacity for instant recognition is hardwired: About half the cells taking information from the visual sensory area to the VWFA—the so-called fast cells—fire rapidly to detect visual changes over a brief time, such as when one scans a page in reading. The content of what one recognizes, however, is not hardwired but learned. Skilled readers learn to recognize letters, words, and numerals in a flash—the way a car enthusiast, for example, learns to identify a vehicle's make, model, and year as it zips past on the highway. In proficient readers and calculators, the VWFA is highly active.[5]

We have been discussing letters and numerals as if the VWFA "thinks" of them as the same sort of symbol. Is reading numerals just like reading letters and words? As discussed in Chapter 5, the answer is mixed. Research has shown that cells sensitive to letters and numerals do not entirely overlap within the VWFA. Some sites within this area respond to numerals as well as letters and words, whereas others respond just to one sort of symbol or the other. When a numerical task becomes too demanding, some letter sites activate to help out. Letter- and numeral-sensitive cells are not inherently different, but according to one theory become trained through one's experience to recognize specific types of symbols and sort themselves into specialized cell neighborhoods through common use and neuronal interaction. For example, one sees mostly letters when reading and mostly numerals when doing math or bookkeeping.[6]

Behavioral studies support the idea that the human brain registers letters and numerals in different categories. For example, in symbol-search tasks, it is typically easier to pick out a letter in a field of numerals than in a field of letters (and vice versa), as illustrated in Figure 6.3.[7] Furthermore, some people with brain injuries that leave them able to read words only letter by letter (e.g., *s-i-x*) have no trouble reading big numerals (*1,340,210*), whereas other individuals with injuries who can read, sequence, and compute with written number words (*six*) cannot do so with written numerals (*6*).[8] Even among healthy individuals, however, letter-numeral discrimination seems to vary considerably from person to person. For example, postal workers accustomed to sorting Canadian postal codes, which consist of mixed letters and numerals (e.g., M5S 1A4), can locate letters in a field of letters just as quickly as they can find them in a field of digits, demonstrating that even adulthood life experience can alter brain function.[9]

These findings raise the possibility that some people with certain reading disorders may have difficulty with numerals as well as letters, whereas others may not.

A

L	C	P	R	G	H	Q	S
C	Q	G	S	R	P	H	L
S	P	R	C	H	Q	L	G
P	H	Q	L	G	R	S	C
C	G	R	P	L	H	Q	S
G	C	Q	L	R	S	P	H

B

7	3	9	R	6	4	2	8
3	2	6	8	R	9	4	7
8	9	R	3	4	2	7	6
9	4	2	7	6	R	8	3
3	6	R	9	7	4	2	8
6	3	2	7	R	8	9	4

Figure 6.3.   Find the letter *R* in A) a field of letters and B) a field of numerals. Which is easier? (*Source:* Polk, Stallcup, Aguirre, Alsop, D'Esposito, Detre, et al., 2002.)

Dyslexia brain studies to date have not focused on numeral processing; it would, of course, be useful to know in general whether certain reading impairments carry a specific risk for math difficulties. Conversely, individuals who read well may struggle with written mathematical language. For instance, researchers reported on a healthy 11-year-old boy with normal phonological skills, counting, and number sequencing; grade-level reading and spelling; and good understanding of numerical place value. However, the boy could not read or write Arabic numerals or even number words.[10] As this example suggests, much remains to be learned about how the brain "reads" quantitative notation.

Naming Numerals

Numerals can be named (*6* ↔ "six") and connected to the verbal counting sequence because of neural links between the VWFA and phonological and speech areas (see Figure 6.2). (There may also be some purely visual numeral–recognition in the right occipital lobe, but it is wired for sound only via neuronal connections to the left-hemisphere phonological center.) These regions, which are interconnected by complex neuronal circuits, serve natural-language functions such as phoneme detection; lexical, morphological, and syntactic comprehension; verbal associations; and verbal memory. Of course, knowing the numerals' names is not hardwired; they must be learned through instruction and practice—much the way we train our brains to attach words to letters, pictures, or objects.

Mathematical language, like natural language, has grammatical properties. Does the brain process the two grammars the same way? The research on this matter—all concerning individuals with injury to different brain sites—has turned up some provocative findings. For example, one man, who made phonemic substitutions when reading nonnumber words (e.g., read *mouse* as "mouth"), instead made lexical errors when reading Arabic numerals and number words (e.g., read *250* as "four hundred and sixty" and *thirty-seven* as "fifty-seven"). Phonemes (e.g., /m/-/ou/-/s/) serve as the fundamental elements of nonnumber words, and reading involves decoding the graphemes (*m-o-u-s-e*) that represent them. This case, by contrast, suggests that digit names (e.g., "six") may form the basic building blocks when the brain reads numerals and even number words. Notably, however, this individual produced a number ("four hundred and sixty") with the same number of digits as *250*, demonstrating that he clearly understood numerical morphology (place value). This error pattern is characteristic of injury to the back of the language circuit.[11]

Other individuals with brain injuries had morphological trouble but were lexically accurate. For example, they wrote "seven thousand forty" as *700040*, or read *602* as "six thousand two," demonstrating that they process numerical morphology independently of digit names. Moreover, these individuals compared numerical values correctly, suggesting that when connecting circuits are traumatically disrupted, the brain processes lexical, morphological, and semantic numerical features independently of one another. People with brain injuries who make morphological errors in reading Arabic numerals typically have damage toward the front of the language circuit, where propositional thought is translated into speech.[12]

Math- and natural-language syntax also operate on separate circuits. Several people with injuries that left them lacking the syntactic skill to interpret *The man killed the lion* versus *The lion killed the man* versus *The man was killed by the lion* still knew that $59 - 13 \neq 13 - 59$ and that $7 \times [4 + 3] \neq [7 \times 4] + 3$.[13] Furthermore, a large study of individuals with either severe syntactic or lexical natural-language impairments due to brain injury revealed that only half of each group had any math difficulties.[14]

It is unknown whether and to what extent these findings apply to students with developmental language disorders; that research has not yet been done. The results do demonstrate, however, the extraordinarily complex circuitry involved in understanding mathematical notation. They show the relative independence not only of natural and mathematical language, but also of their various components. In the classroom, this means that students with natural-language or reading difficulties may— or may not—also have trouble learning fundamental number skills. Only careful evaluation can determine the extent of such co-occurrence for any individual student.

Determining Numerical Value

How does the brain attach value to numerals and number words—the semantic feature of mathematical language? Neural circuits link the VWFA and phonological regions to the IPS, which register digits both independently and by place value in multidigit numerals.[15] This function is not hardwired. Just as our visual system can be trained through instruction and practice to read numerals, so can these quantity-sensitive IPS regions be taught to endow numerals and number words with quantitative meaning. (Neuroscientists are currently exploring the possibility that the left and right IPS are different—for example, that perhaps the left IPS is more sensitive to written and spoken numbers than the right IPS.[16])

By adulthood, the circuits connecting the various mental representations of number are well established; most adults can comprehend the value of both numerals ($6 \leftrightarrow$ '—'—'—'—'—'—*) and spoken number words ("six" $\leftrightarrow$ '—'—'—'—'—'—*) automatically. The orthographic association (e.g., *six* $\leftrightarrow$ '—'—'—'—'—'—*) is also automatic for most adults, although it is slower than the others. This finding suggests that one may first have to mentally transcode *six* to "six" or *6* to get its meaning— hence some people's awkwardness in reading dates, times, and monetary amounts written out in number words.[17]

For most people, the three key components—visual numeral and word recognition, phonological number naming, and quantity comprehension—work together seamlessly, just as the brain effortlessly connects the written words to their sounds and meanings when people read fluently. When any of the links develop poorly or are disrupted, however, the brain must make do with the skills that remain intact. For example, because of traumatic injury to the phonological center or to the fibers connecting it to the VWFA, some people cannot read *67* or *89* aloud or write them to dictation; they can figure out, however, that *89* has a higher value than *67* because of the direct links between the visual centers and the IPS in either hemisphere, as shown in Figure 6.2.[18] (The brain uses a similar mechanism for traumatically disrupted natural language: Individuals with intact right hemispheres can sometimes still classify words, even though they cannot read the words aloud.)

One individual with injured language areas of the brain, who could not read or spell nonnumber words (e.g., *cat*) and had lost much of his general knowledge, was still able to read number words (e.g., *six*) and numerals. Although he could not write the arithmetic operation signs, he could still compute, estimate, and correctly place numerals on a number line. How did he do that? One must suppose that he drew on his remaining intact connections; for example, he might have gone from *six* to "six" as follows: *six* → 6 → '—'—'—'—'—'—* → "six" or *six* → '—'—'—'—'—'—* → 6 → "six".[19] People sometimes find remarkable ways to compensate for lost skills.

Arithmetic Facts and Calculation

Another complicated issue has to do with the way the human brain transforms numbers. Several questions have guided the neuroscientific research: Does the brain handle exact and approximate calculations alike? Does it process all simple arithmetic operations the same way? To what extent does it rely on natural language and reading circuits to store and retrieve math facts? This section addresses these questions.

As illustrated in Figure 6.4, calculation recruits the brain's left parietal lobe at the intersection of all the math-language functions shown in Figure 6.2. Finger awareness also resides in this region, supporting the notion that fingers do, indeed, play an important role in calculations and one's understanding of number.[20]

Exact versus Approximate Calculations

Although both hemispheres are capable of approximate calculation, "split-brain" and brain injury research has demonstrated that exact calculation, like exact numeral reading, depends on the left hemisphere. Without the normal input from the left-hemisphere phonological centers, the right hemisphere's calculations are only approximate. The right hemisphere can verify that $2 + 2 \neq 9$, but is less sure about $2 + 2 \neq 5$, revealing the influence of Weber's law, as discussed in Chapter 2.[21] Moreover, although finger awareness resides in both hemispheres, it is directly linked to counting and calculation only on the left.[22]

In intact brains, the hemispheres communicate with each other and both contribute to calculation. Even when the interhemispheric circuits are working, however, the left hemisphere calculates more quickly and accurately than the right. For complex calculations, the brain recruits both exact and approximate capabilities in tandem.[23]

Simple Calculations

Do students use the same parts of their brains for addition, subtraction, multiplication, and division? Not entirely: The brain activates in somewhat different sites, depending on the arithmetic task. The clearest distinction is between subtraction and multiplication. Single-digit subtraction, as well as the related function of number comparison, tends to activate the IPS on either side of the brain, in keeping with the

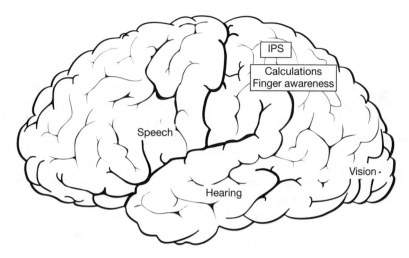

Figure 6.4. Calculation and finger awareness in the brain's left hemisphere. (*Key:* IPS = intraparietal sulcus.) (*Sources:* Duffau, Denvil, Lopes, Gasparini, Cohen, Capelle, et al., 2002; Roux, Boetto, Sacko, Chollet, & Tremoulet, 2003; Rusconi, Walsh, & Butterworth, 2005.)

view that those operations are relatively easy to visualize along the mental number line. Simple multiplication, on the other hand, is largely confined to the language centers of the left hemisphere, in line with the idea that people rely on written and spoken language for that function because multiplication is difficult to envision on the mental number line. (The physical number line is an excellent tool for teaching multiplication concepts, but creates a poor template for mental multiplication manipulations because the numbers quickly become unwieldy.) Behavioral research has found that learning leads to more efficient solution strategies in subtraction but to more automatic retrieval in multiplication, in line with their neuronal distinction. Addition and division, which are amenable to several learning and recall strategies, activate multiple overlapping regions.[24]

These findings come to us from functional imaging studies, as well as from studies of individuals with brain injuries. Information is also provided by a study of one woman, whose arithmetic functions in her left parietal lobe were mapped during surgery to remove a brain tumor there. To make sure they did not unnecessarily remove portions of her brain needed for language and arithmetic, doctors tested her during surgery, which they performed under local anesthesia so that she could respond. Doctors used pinpoint electrical stimulation of the exposed brain, which prevents the affected cells from performing their usual function for the duration of the stimulation. By testing her skills before and during stimulation, they mapped her parietal lobe. The tasks, which were all oral, included (among others) subtracting serially by 7 and stating low-value multiplication facts.

The researchers located different distinct neighboring sites for subtraction and rote multiplication in the lower portion of the left parietal lobe, with a buffer area devoted to both operations. As Figure 6.4 illustrates, the IPS lies along this region's upper edge. Notably, subtraction activated the area directly abutting the IPS, linking the operation closely with the mental number line, while multiplication activated an area away from the IPS, in the center of the region essential to language and reading.[25] Does this explain how some people can multiply numbers without necessarily thinking about the values they signify? Some investigators think so.[26] (For the concerned reader: The surgery was successful. The surgeons completely removed the tumor and the woman retained her arithmetic skills.)

That the area governing language and reading activates during simple multiplication, regardless of the problem format, suggests that the brain codes the facts phonologically (e.g., "six times four equals twenty-four")—a notion supported by the tendency of bilingual students to recall arithmetic facts most efficiently in the language in which they had learned them (see Chapter 7). Other evidence, however, indicates that the brain may store math facts as visual images (*6 ×4 = 24*) rather than as phonological sequences.[27] In fact, scholars debate whether the brain stores multiplication facts in a single code, several codes, or even several interacting codes, depending perhaps on the task at hand, individual preference, or the way one had been taught.

Although these patterns are difficult to detect in large-scale studies, one could resolve this code question by teaching some facts verbally (by class recitation, as is done in some settings), some numerically (with flash cards or on a computer monitor), and even some in written words, and then by testing each fact in all the formats. This experiment would be difficult to conduct in the classroom, of course, but one researcher had the opportunity to examine and retrain an individual who had suffered a brain injury that had, oddly enough, robbed him of only his multiplication facts. The individual could still read, write, match, and compare numerals, and he remembered the algorithm for complex multiplication. The investigator retaught him each difficult fact (not the easy zeroes, ones, fives, or squares) in only one format: for example, *2 × 3 = 6, seven times six equals forty-two*, or "four times three equals twelve." Following training, the individual performed each fact flawlessly in all formats, but he performed them fastest in their trained format, suggesting that how people code and recall number facts may depend to some extent on the format in which they learned them.[28]

Regardless of how the brain codes arithmetic facts, neuroscientists know from studying individuals with brain injury that the brain stores them separately from other facts, including other kinds of numerical information such as the boiling point of water, one's birthday, or even arithmetic signs, operations, and properties.[29] One may speculate that this at least partly explains why some students have little trouble remembering material from history or science class but cannot remember the multiplication table.

Complex Calculations

Technical limitations make it difficult to study brain function during more complex calculations. Broadly speaking, however, the more complex the computation, the more diverse the recruited brain regions—including the IPS in both hemispheres, language regions, and the working memory areas to be discussed shortly.[30] These findings support the notion that people draw on many different strategies using many different formats—visual and auditory fact recall, quantity manipulation, comparison, and others—to solve complicated calculations. Other studies have demonstrated that different regions activate depending on whether a person is thinking of an algebra problem in story form, equation format, or as a diagram.[31] The evidence also implies that computational difficulties can derive from many possible sources.

The foregoing discussion highlighted the complex nature of the mathematical-language cerebral networks. Many questions remain unresolved, however. For example, some individuals with autism and poor language skills have extraordinary computational ability.[32] Is it possible that for some people, math- and natural-language circuits are largely independent? Evidence also suggests that the brain codes and executes mathematics in a much more specific, modular way than is yet understood. The untested possibility that the mathematical brain circuitry is arranged in parallel rather than in series—where one burned-out bulb does not short circuit all the others—would not only help explain preserved capabilities among individuals with traumatic brain injury, but may also provide insight into the nature of developmental mathematical difficulties. Clearly, much more research into the neurology of mathematics is warranted.

Developmental Changes in the Child's Brain

Most mathematical neuroscience research has focused on adults. From the educational perspective, however, the important questions have to do with children's number-skill acquisition and whether their performance is linked to maturation in brain function. A handful of important preliminary studies have addressed this issue.[33]

Much of adult brain activity during conventional math tasks occurs in the back part of the brain, particularly in the IPS and the language and reading regions of the parietal and occipital lobes. This is not the case with children: Much of the action in young children's brains while they are doing symbolic, as well as nonsymbolic, math occurs in the frontal lobe, in a region dedicated to such general functions as working memory and attention. Psychologists speculate that children rely on these general skills while the highly specific numerical abilities of the rest of the brain slowly mature.[34]

Children demonstrate a similar developmental pattern when they learn to read. To become a fluent reader, it takes several years to train the brain's visual centers in a particular writing system. Indeed, studies have shown relatively little activation in the VWFA during the early stages not only of reading, but also of arithmetic acquisition, with increasing activity as children learn to read words and numerals fluently.[35] In addition to the VWFA, areas that show increased activation with age during mathematical tasks include portions of the IPS and the left parietal cortex. For math, this gradual rearward and leftward shift can take many years and much hard work, well into adolescence.[36] Brain activation shifts in a similar way among adolescents and adults as they acquire new math expertise, moving from large, general sites controlling effortful working memory and quantitative manipulation to more focal regions of the left parietal cortex and iconic visual pathways.[37] In this way, learning alters brain circuits, making them more efficient and reducing the amount of mental energy required for a task.

How children learn to recognize correct digit orientation remains unresolved. The iconic visual pathway inherently generalizes across the vertical axis so that similar numerals are easily confused; when children learn to read, they must "unlearn" this natural equivalence in regard to asymmetric letters, numerals, punctuation, and mathematical symbols, as discussed in Chapter 5. It is not yet known how the brain resets the default option, although both spatial and iconic visual systems seem to be involved. Once trained to recognize symbols only in their correct orientation, however, the iconic pathway can no longer read them backwards; the brain must rely on the slower mental rotation capacity of the spatial circuits to do the job. When the spatial pathway of an older child or adult is injured, reading in the mirror becomes impossible.[38]

Is it possible that some brains are especially well equipped for formal mathematics from the beginning? At present, very little information exists on this subject. Postmortem brain studies of several prominent mathematicians, including Albert Einstein and Carl Friedrich Gauss, have revealed enhanced growth of the parietal lobes, particularly on the left, as well as of the fibers connecting the hemispheres.[39] Were they born that way or did those regions develop from use? The scientists who examined Einstein's brain concluded from the growth pattern that the extensive development occurred early. More research—using less-invasive techniques—should help to clarify this important question.

Very little is known about numerical processing in the brains of children who struggle with math. Considerably more is known about reading disorders. For example, children with dyslexia demonstrate weak VWFA activity during reading, past the age when that region normally becomes active; indeed, they tend to read slowly and some continue to write letters and numerals in mirror image as they mature. Sometimes these children are also slow at naming numerals and letters. One theory attributes these difficulties to an abnormality of the fast cells that transmit information at the rapid rate necessary for reading. Such children frequently show increased activation in frontal and right hemisphere regions as they struggle to compensate.[40] One may speculate that this pattern also pertains to some children with developmental math disabilities, with and perhaps without reading difficulties, but there is no scientific evidence yet to support that theory.

To date, three small studies have examined brain structure or function in children with isolated math difficulties and have found abnormalities in the IPS. One of those studies also revealed abnormalities in brain regions responsible for spatial memory and fact retrieval, as well as in the frontal areas governing the attention and working memory that young children rely on as they learn math.[41] More broadly, among children and adolescents with both poor math performance *and* signs of cognitive difficulties, left-hemisphere-type impairments tend to compromise arithmetic acquisition more than do right-hemisphere-type impairments.[42] A program of functional imaging studies relating the location and intensity of cortical activation to the fluency and accuracy of children's performance on various math tasks should provide the kind of useful information that is already available for reading acquisition and dyslexia. Moreover, dyslexia research has shown that specialized reading instruction can normalize the reading circuit, demonstrating that experience can actually alter brain function.[43] One looks forward to the day when this kind of scientifically based pedagogy is available for math intervention as well.

Classroom Implications

The neuroscientific findings reviewed here bolster conclusions drawn from behavioral studies—namely, that symbolic mathematics overlaps very imperfectly with natural language and reading in the brain. Although some scholars have posited a developmental stage in which young children establish general symbol systems for expressing, remembering, and communicating information,[44] the research suggests that alphabetic and numeric systems are at least partly independent, although they do draw on many of the same underlying perceptual and cognitive resources. In the classroom, this means that some children with language and reading disorders may have difficulty with symbolic mathematics, including mastering basic arithmetic facts, whereas others may not. Conversely, some children who read well may struggle with formal mathematics. Of course, children may find mathematics difficult for reasons other than language.

These studies also lend support to the idea that explicit instruction and practice facilitate learning for some students. In one study of college students, researchers found that those with math disability had difficulty associating numerals with their values ($6 \leftrightarrow$ '—'—'—'—'—'—*). They discovered, however, that when teachers explicitly directed students' attention to the quantity represented by each numeral, the students became more proficient and fluent.[45] Most students develop this proficiency naturally in the primary grades; others do not catch on without explicit help. Similarly, students may require explicit instruction to learn numerical shapes, names, and even finger associations. Future research must continue to explore which pedagogical techniques, if any, enhance the development of mature brain activation patterns.

Some studies suggest that people code arithmetic facts phonologically; others point to nonphonological—perhaps visual—coding. Multisensory pedagogical theory suggests that students learn best when provided with the same information in multiple formats. This theory requires rigorous testing in relation to learning math facts. In particular, research should determine whether children with particular cognitive impairments benefit from one format or another. For example, would children who have trouble reading numerals learn math facts more easily if they were drilled orally through recitation or song rather than with flash cards and written or computer exercises? Would drilling facts both ways simultaneously enhance their symbol reading fluency? Many questions remain unanswered.

Finally, training—explicit instruction, exposure, drill, practice—changes brain function and even brain structure. Thus pedagogy has implications for research, yet few neurocognitive studies account for how their participants learned and encoded the facts in the first place.[46] Had they drilled by recitation? Flash cards? Computer? Writing out the problems? Much of the literature on brain function in mathematics has come from European countries with a strong tradition of oral recitation math-fact drill. Do those studies' conclusions apply to individuals who learned visually by flash cards, computer drill, or written exercises, or even by strategy analysis?

Students tend to find the methods that work best for them regardless of their teachers' intentions. More needs to be learned about the relationship between teachers' goals and actual student behavior. Research comparing brain function across modalities and problem-solving strategies will have significant implications for the classroom. Similar studies focusing on math-task–related brain functioning of students across the ability spectrum in language, reading, and math would be especially welcome.

Conclusion

Conventional mathematics is founded on mathematical language. Research using exciting, new technologies has revealed that the neural networks supporting mathematical language are extraordinarily complex. Like a machine with multiple moving parts, these circuits are vulnerable to disruption at many points, supporting the idea that children can have trouble with math for diverse reasons.

Moreover, this mathematical "tongue," like natural language, is both spoken and written, and learning it demands many of the same perceptual, cognitive, and motor skills. It is not yet clear, however, whether and to what extent the networks supporting mathematical and natural language overlap. Letters and words are not the same as number words, numerals, or other mathematical symbols. Moreover, not all children with language and reading disorders struggle with math, and not all those with math difficulties perform poorly in other areas.

Cognitive science has made significant strides in understanding reading acquisition and the nature of dyslexia. It has not focused as much attention on mathematical learning and disability, and its findings on mathematical cognition, though promising, are not yet directly applicable to pedagogy. Its insights should be useful to educators and psychologists, however, in their efforts to understand why some students struggle with math. Fortunately, neurocognitive work in mathematical cognition has accelerated, and research can be expected to close the gap with reading in the coming years.

Notes

[1] See Tang et al., 2006.

[2] See Dehaene, Molko, Cohen, & Wilson, 2004; Dehaene, Piazza, Pinel, & Cohen, 2003; Rusconi, Walsh, & Butterworth, 2005.

[3] Gazzaniga, 1995.

[4] See, e.g., Casey & Brabeck, 1989; O'Boyle, Benbow, & Alexander, 1995.

[5] Cohen & Dehaene, 2004; Simos et al., 2002; Stein, 2001.

[6] Allison, McCarthy, Nobre, Puce, & Belger, 1994; Dehaene, 2005; Polk et al., 2002.

[7] Jonides & Gleitman, 1972, as cited in Polk et al., 2002.

[8] See, e.g., Cipolotti, Warrington, & Butterworth, 1995; Warrington & Shallice, 1980.

[9] Polk & Farah, 1995.

[10] Temple, 1989.

[11] Cohen, Verstichel, & Dehaene, 1997.

[12] Delazer, Girelli, Semenza, & Denes, 1999; McCloskey & Caramazza, 1987.

[13] Varley, Klessinger, Romanowski, & Siegal, 2005.

[14] Basso, Burgio, & Caporali, 2000.

[15] Fias, Lammertyn, Reynvoet, Dupont, & Orban, 2003.

[16] Ansari, 2007.

[17] Dehaene, Bossini, & Giraux, 1993; Eger, Sterzer, Russ, Giraud, & Kleinschmidt, 2003.

[18] See, e.g., Cohen & Dehaene, 1996; Rossor, Warrington, & Cipolotti, 1995.

[19] Butterworth, Cappelletti, & Kopelman, 2001; B. Butterworth, personal communication, January 17, 2005.

[20] Roux, Boetto, Sacko, Chollet, & Tremoulet, 2003; Zago et al., 2001.

[21] See Dehaene & Cohen, 1991.

[22] Roux et al., 2003; Rusconi et al., 2005.

[23] Dehaene, Spelke, Pinel, Stanescu, & Tsivkin, 1999; Stanescu-Cosson et al., 2000; Venkatraman, Soon, Chee, & Ansari, 2006.

[24] Cohen, Dehaene, Chochon, Lehéricy, & Naccache, 2000; Dehaene et al., 2003; Grabner et al., 2007; Ischebeck et al., 2006; Lee, 2000; Lemer, Dehaene, Spelke, & Cohen, 2003; Tohgi et al., 1995.

[25] Duffau et al., 2002.

[26] Chochon, Cohen, van de Moortele, & Dehaene, 1999.

[27] Rickard et al., 2000; Varley et al., 2005.

[28] Whetstone, 1998.

[29] See, e.g., Cappelletti, Butterworth, & Kopelman, 2001; Ferro & Botelho, 1980; Hittmair-Delazer, Semenza, & Denes, 1994; Warrington, 1982.

[30] See, e.g., Gruber, Indefrey, Steinmetz, & Kleinschmidt, 2001; Kong et al., 2005; Menon, Rivera, White, Glover, & Reiss, 2000; Zago et al., 2001.

[31]Lee et al., 2007; Sohn et al., 2004.

[32]Hermelin & O'Connor, 1990.

[33]See Ansari, 2008.

[34]Ansari & Dhital, 2006; Ansari, Garcia, Lucas, Hamon, & Dhital, 2005; Kaufmann et al., 2006; Rivera, Reiss, Eckert, & Menon, 2005.

[35]Cohen & Dehaene, 2004; Turkeltaub, Gareau, Flowers, Zeffiro, & Eden, 2003.

[36]See, e.g., Rivera et al., 2005.

[37]Ischebeck, Zamarian, Egger, Schocke, & Delazer, 2007; Qin et al., 2004.

[38]Vinckier et al., 2006; Walsh & Butler, 1996. But see Vingerhoets et al., 2001.

[39]Witelson, Kigar, & Harvey, 1999.

[40]Meyler, Keller, Cherkassky, Gabrieli, & Just, 2008; Shaywitz et al., 2002; Stein, 2001; Wolf, Bowers, & Biddle, 2000.

[41]Levy, Reis, & Grafman, 1999; Price, Holloway, Räsänen, Vesterinen, & Ansari, 2007; Rotzer et al., 2008.

[42]Shalev, Manor, Amir, Wertman-Elad, & Gross-Tsur, 1995.

[43]Meyler et al., 2008; Shaywitz et al., 2004.

[44]See Munn, 1998.

[45]Rubinsten & Henik, 2005.

[46]See, e.g., Delazer et al., 1999; Zago et al., 2001.

Chapter 7

More Sharks in the Mathematical Waters

S o far, we have discussed some of the difficulties frequently encountered in acquiring the basic tools of formal mathematics: counting, arithmetic facts, and simple calculations. This chapter considers several other aspects of conventional mathematics—from the everyday vocabulary of the mathematics classroom to the mysteries of fractions—that even good math students often find confusing or counterintuitive. The scientific research in some of these areas is considerably sparser than it is for basic number skills, but we draw on the existing work to elucidate some of the difficulties inherent in these vexatious topics and to encourage future research.

Vocabulary

> "Reeling and Writhing, of course, to begin with," the Mock Turtle replied, "and the different branches of Arithmetic—Ambition, Distraction, Uglification, and Derision."
>
> From *Alice's Adventures in Wonderland,* by Lewis Carroll[1]

Formal mathematics, like natural language, has a vocabulary. Generally speaking, children's earliest natural language consists primarily of short, simple words that make only basic demands on auditory discrimination, articulation, and memory. These words relate to a young child's familiar experience. Likewise, children master the counting words—also short and easily learned as a sequence—in the context of salient, everyday activity. In both cases, the experiences and the words reinforce each other and together elevate the child's understanding.

Many math words are more complicated than the early counting numbers, however. English-speaking children often get stuck on counting numbers greater than 10 because the number-word forms for these unfamiliar quantities are irregular, for example. Other early grade-school math words are complex: *arithmetic, addition, subtraction,* and *multiplication,* as well as such contemporary regulars as *decomposition* and *manipulatives.* To learn to identify, pronounce, or read these multisyllabic words, children must possess good phonemic awareness (i.e., be able to discriminate the different sounds of each word), lest they wind up as hopelessly confused as the Mock

Turtle. Children with language disorders and many of those with dyslexia have particular difficulty with those skills.[2] One large-scale study of children with dyslexia found that a weak grasp of math terms may have played a role in their mathematical difficulties.[3]

Not all math words are unwieldy: Several of the earliest math words students learn—*number, add, divide, sign, carry, borrow, trade*—are short and have a familiar related use in everyday life, at least for most children. Others—*plus, minus, equal, sum, zero*—are technical terms specific to mathematics and, although they take on derivative, nonmath meaning for older children and adults, carry no intrinsic clue about their meaning for young students. Meanwhile, other words may no longer be as familiar to children as they once were. For example, children of the digital age may not know what *clockwise* means. It is difficult to learn a new concept when it is explained with only an unfamiliar technical term; some common, familiar language is required to link concept and terminology. If both the word and the concept are utterly unfamiliar, the child may be left in the dark on both counts. Vocabulary can impede conceptual development, as well as enhance it.[4]

Some math words do have helpfully familiar common meanings, but with confusing usage rules or ambiguous connotations. For example, certain magnitude words apply to both continuous and discrete amounts (*more milk, more cookies*). Others apply to only one (*there is little time and even less money*) or the other (*he has few apples and many pears*), although *less than* always applies to amounts of time, money, and distance (*he has less than $30*). Grammar authorities are silent on the correct use of *less than* and *fewer than* with fractional amounts (*less/fewer than one third of the cookies, less/fewer than two thirds of the milk*) and percentages. Math uses two early-learned words, *right* (signifying correctness, an angle, and a direction) and *left* (meaning remaining, departed, and a direction) in most of their senses. Thus students may wonder: Is a right angle one that opens to the right? Is an angle that opens to the left a left angle? How can a left angle be a right angle? Finally, the technical meanings of other math words are only marginally related, if at all, to their common meanings: *leg* and *face, rational* and *irrational, power* and *exponent, root* and *radical*. And what is so improper about an improper fraction? Or odd about an odd number?

Children often find the arithmetic operation words especially confusing. Children must learn several words to master addition (*add, plus, sum, combine, altogether, increase, grow, more*) and a similar list for subtraction. *Sum* is easily confused with *some*, itself a quantity word; a student might become baffled when asked "What is the sum?" but can easily "Add this to that." (For that matter, even simple number words have homophones: *to, too, for, ate*. Children may wonder what to make of "Do I have to count this one, too? What for?"[5]) *Subtract* and *minus* give no clue to someone who does not know Latin about the arithmetic transformations to which they refer; if a child does not understand the operation, the words will offer no help. The more common terms *from* and *take away* are ambiguous as to which value remains behind; indeed, one instructor promotes the term *punch away* to emphasize which amount the child should discard,[6] whereas others recommended a simple *less*.[7] *Less*, however, also denotes relative value; the children in one Australian study concretely associated *less* with subtraction simply because in whole-number subtraction the second number is always "less" than the first one. Those children also confused the subtraction concept *difference* with *different* and thus did not associate it with arithmetic at all. For example, one "difference" between 7 and 4 is that 7 is odd and 4 is even. Furthermore, classroom demonstrations apparently led them to link the subtraction term *compare* with addition instead, as in "When you compare groups you put them in pairs . . . you join them."[8] *Multiply*, whose etymology signifies augmentation, could logically refer to both addition and multiplication; augmentation does not apply to the multiplication of proper fractions, however, in which *divide* augments instead, or to the product of a positive and negative number. *Product*, which connotes something produced, could reasonably refer to any arithmetic transformation.

The various technical words for the arithmetic operands (*augend, addend, sum, minuend, subtrahend, difference, multiplier, multiplicand, product, dividend, divisor,* and *quotient*) require linguistic sophistication and frequent use to become part of a child's working mathematical vocabulary. As to *terms* and *factors*, the words themselves do not give a clue that one has to do with addition and subtraction, whereas the other pertains to multiplication; failure to distinguish them can lead to serious algebra difficulties. Even in the upper grades, most students learn words such as *numerator* and *denominator* strictly by rote and must stop to think when asked about the *least common multiple*, the *greatest common factor*, or the *lowest common denominator* (where *shared* might work better than *common*). And how are they to understand that to reduce a fraction leaves it the same?[9]

Other terms suffer from syntactic flaws. For example, *times* works well in the phrase *four taken three times*, but is confusing in its usual syntax, *four times three*. Why not simply *four threes*? Or even *groups of*, as in *four groups of three*. One might reasonably wonder why *times* and *by* are used for whole number multiplication, but *of* is used for fractions. Math uses the preposition *by* to denote quantitative relationships in subtraction (*three exceeds two by one*, or *eight differs from five by three*) but to signify, in passive voice, actual transformations in multiplication (*two multiplied by three*) and division (*ten divided by two*). *Divided by* evokes the White Queen's image of "a loaf divided by a knife," as cited in the book's introduction; *divided into* might work better, as in *twelve divided into threes* or *into three parts*. *Goes into* has nothing to do with walking through the classroom door, the more familiar meaning of the phrase; indeed, some teachers have dubbed the long-division radical $\overline{)}$ the *guzinta*—as in, "four guzinta twelve equals three"—which makes just as little sense as *goes into*.[10] Others have suggested *contained in*, as in *four is contained in twelve three times*.[11] Little prepositions are often taken for granted, yet in mathematical language, they are both significant and problematic. In fact, one teacher has documented how prepositional differences in African American vernacular can have a crippling effect in a math classroom that uses Standard American English.[12]

Mathematical language is different from ordinary conversational or most literary language in that it is terse and compact; as one observer noted, it is more like income tax forms or poetry than normal conversation.[13] (Some of the shortest college textbooks are for advanced math classes.) For this reason, math text is tough to read, even when the vocabulary is simple. Each word is highly specific and carries a great deal of weight. For example, one might say, "John stopped *by* the house the other day," meaning John visited and probably came *into* the house. In math, however, *twelve divided by six* means something very different from *twelve goes into six*. Properly written, math leaves no room for connotation or interpretation.[14] When the vocabulary is unclear, however, many students are left to their own interpretative devices, making their job even harder.

So what's a teacher to do? The educator has several potentially conflicting jobs: to teach the mathematical concepts, to impart commonly and historically shared language, to use technically accurate terms, to communicate comprehensibly with students, and to teach children how to communicate mathematically themselves. Few mathematical terms used in English suit all pedagogical purposes. Other particular languages have more straightforward mathematical vocabulary. In Chinese, for example, *triangle* translates to *three-cornered shape*, an easier reach for children who cannot yet parse *tri-angle*.[15] English-language literacy specialists recommend incorporating etymology into vocabulary instruction.[16] With English math terms, however, the historical roots are sometimes enlightening and sometimes not, particularly for the youngest students.

Rewriting the dictionary, while an appealing option, would sever ties with history and tradition and alienate parents. Some educators suggest not only introducing mathematical ideas first by hands-on experiments and demonstrations, but also encouraging students to explain those insights in their own language. Once the students understand the concept in their own words, the teacher can instruct them about the proper terms.[17] This idea has not yet been tested, however. Another educator found that simply teaching math vocabulary explicitly and systematically (eight words a week for 30 minutes), as one would nonmath terms, significantly improved his fifth graders' understanding and problem solving.[18] Children with dyslexia and language disorders may especially benefit from such instruction.[19] When several terms are available for a single concept, however, the teacher must still make choices—for example, between the technically accurate *decompose/recompose*, the more familiar and accessible *trade*, or the *borrow/carry* the students' parents and grandparents know and can help them with.

Finally, math word problems introduce broader vocabulary issues. Often children perform poorly on math tests, not because they do not understand the math concepts, but because they do not comprehend the problems' general nonmath vocabulary. This difficulty plagues not only children with language and reading impairments, but also students from language-impoverished environments and those who are learning math in their second language. Researchers recently reexamined the wording used in some U.S. national math tests to ensure that the tests are assessing knowledge of math and not, inadvertently, knowledge of English vocabulary or even of unfamiliar aspects of American life. In one large-scale study, all students performed better when examiners provided them with a glossary and extra time, but only simplifying the language in which the questions were written narrowed the score difference between students with limited English skills and those who were proficient.[20]

The vocabulary of the math classroom—both technical and general—can impede, rather than enhance, math learning. For educators struggling to illustrate complex mathematical ideas and procedures on the chalkboard, it is easy to overlook this basic verbal component of classroom communication. The math vocabulary conundrum is wide open for research.

How Many More . . . Than?

Word problems have troubled math students since math class was invented. Of all the various sorts of basic word problems, however, the most deceptively simple one asks the student to use arithmetic to compare values (e.g., "John has four marbles. Mary has six marbles. How many more marbles does Mary have than John?"). Problems containing the phrase *how many more . . . than* and its variations are quirky and stand out among word problems as being particularly vexing for students and puzzling from a cognitive point of view.

What makes this sort of problem so devilishly difficult? Clearly the trouble is not arithmetic. Even young children can easily compare two quantities, and children who struggle with these problems can usually perform the basic computations. Nor is vocabulary the culprit: Preschoolers know what the words *how, many, more,* and *than* mean, or at least can use them each correctly; the words found in the problems' variants—*fewer* or *times,* for example—are equally familiar.

The difficulty lies in the verbal construction *how many more . . . than.* In one study, cited previously, a researcher asked a group of young children the question, "There are five birds and three worms. How many more birds are there than worms?" Only 17% of the preschoolers and 64% of the first graders answered the problem correctly, even when the investigator visually displayed the information and highlighted the one-to-one correspondence. Then he rephrased the second line: "Suppose the birds all race over and each one tries to get a worm! How many birds won't get a worm?" With the question restated, 83% of the preschoolers and all of the first graders responded correctly.[21] Even 8- and 9-year-old students sometimes vaguely associate *many more* with addition (e.g., "When you add you get a lot more—many more") or conflate it with their understanding of borrowing in subtraction ("When you have 54 and take away 29, you can always go and get many more" [by borrowing]).[22] The original wording, in deceptively simple English, served to obscure rather than clarify the crux of the mathematical problem, leading to difficulty in determining the appropriate mathematical model for it. How to set up a problem is a common dilemma for many other types of word problems, but is especially difficult in comparisons.

Children come to understand comparative verbal constructions only gradually as they mature. Another study tracked this developmental trajectory into high school. Second graders, who could read the problems themselves, had no trouble answering, "Which is more, 10 or 13?" or "Which is less, 7 or 9?" It was not until fourth grade, well after they had learned to add and subtract single digits, that children could reliably answer "What number is 1 more than 5?" and "What number is 2 less than 5?" Note that the format of these questions is direct; that is, the operations needed to solve them match the words describing the relationship, such that *more* signals addition and *less* cues subtraction. Harder still is the indirect form of the question—"The number 8 is 2 more than what number?" and "The number 6 is 2 less than what number?"—because it requires the incongruent operation (in this case, *more* indicates subtraction and *less* indicates addition). Although this study found that students could answer reliably around sixth grade, another study found college students still struggling with these problems. When baffled students answer comparison questions incorrectly, they frequently either just offer one or both of the numbers already stated in the problems (as if answering the simpler questions, "How many . . . ?" or "How many is the set that has more?") or, in the case of indirect phrasing, use the congruent rather than correct incongruent operation.[23]

There are many other quantitative comparisons beside *more, less,* and *fewer,* including *longer/shorter, heavier/lighter,* and so forth. Linguistically, children usually master these other constructions even later than *more* and *less.* The point, however, is that solving any of these arithmetic comparisons involves not just mathematical acumen, but linguistic sophistication and maturity. Children with poor English skills, for whatever reason, find these verbal constructions particularly tough. One study found that the children from Papua New Guinea (whose initial contact with English, or indeed with any written language or quantitative terms, is on the first day of school) lagged several years behind their peers who are native English speakers in solving comparison problems. They still could not make indirect arithmetic comparisons reliably in 10th grade, the oldest grade studied.[24]

Is it possible to teach children how to make sense of these problems? Several investigators found that diagramming helped both second graders and college students learn how to translate these peculiar and difficult verbal constructs into useful mathematical statements—a strategy that worked better than sentence-by-sentence linguistic analysis.[25] More research is warranted.

Bilingual Math

Certain aspects of math—such as counting, arithmetic facts, algorithms, and exact calculations—draw heavily on broadly defined linguistic skills, whereas others— such as approximation, certain numerical manipulations, and geometric reasoning—recruit chiefly nonverbal skills. If this cognitive categorization is accurate, psychologists have reasoned, then people should transfer some math skills more easily than others to a new natural language.

To test this idea, researchers examined a small group of well-educated Russians who had learned English as adolescents and young adults. In the experiments, the investigators taught these bilingual individuals mathematical information in English or in Russian, using written words, and tested them on the lesson in both languages. They also taught them fictional geography and history lessons containing numerical information. The researchers found that the participants efficiently retrieved new *approximate* numerical information (e.g., estimations of complex sums and cube roots), regardless of the language in which they were trained or tested, and easily generalized the knowledge to similar problems presented in either language. The language of training was also irrelevant for nonnumerical information gleaned from the lessons.

On most other number tasks, how efficiently the participants retrieved information depended on the match between language of training and that of testing. They retrieved freshly learned *exact* numerical information, such as complex sums and number facts derived from stories (e.g., number of kings, age of a character), more efficiently in the language in which they had been taught it. The only exception to this pattern was multiplication, for which both exact and approximate facts were retrieved more efficiently in the language in which they had been taught, probably because one can access multiplication approximations only through preliminary exact computations. Participants strongly preferred to compute in their native Russian language, however—a common sentiment among bilingual people. They also retrieved some information from the geography and history lessons (e.g., small-number fractions, ordinals, time of day, spatial directions) more readily in Russian, even if they had been taught the lessons in English, perhaps because the vocabulary of these elementary concepts was deeply engrained in Russian.[26]

These findings suggest that exact number and calculation (particularly multiplication) depend on language but that approximate quantitative thinking does not—notions that concur with those previously seen about both how young children acquire mathematical skills and how the brain encodes numerical information. Furthermore, the mental representation of exact mathematical information, at least when it is taught in verbal format, appears to rely on the specific natural language in which the individual acquired it. As early as preschool, linguistic differences, such as the structure of plurals, can affect children's number concepts differently across languages.[27]

If researchers replicate these general findings in bilingual schoolchildren, the results will have important implications for their education. For example, in doing exact calculations, remembering a wide variety of number facts, or even counting or using number words, such children may access the information learned in their first language differently or separately from knowledge acquired in their second. (Except for individuals who are perfectly fluent in both languages, brain circuitry is slightly different for a second natural language than it is for the mother tongue.[28]) This linguistic disconnect could potentially lead to inefficient performance or confusion, especially when the conceptual emphases of the two languages differ. For example, Koreans name the denominator before the numerator in identifying a fraction, emphasizing the total number of equal-size parts, whereas English speakers do the reverse. Thus, in one small study of bilingual Korean American children, when the students explained fractions in English, they wrote the numerator first; when they explained them in Korean, they wrote the denominator first. Although they found it much easier to explain their work in Korean, they knew the technical terms, such as *common denominator,* only in English.[29]

Subtle cultural differences also exist in the Arabic notational system, which is now used nearly globally in education and commerce. For example, some Japanese fractional expressions include Kanji and Hiragana characters. Moreover, Japanese and Koreans never use a slash mark (/) for fractions or division and their not-equal sign reverses the direction of the slash ($\neq$); other cultures use a comma instead of a period to separate whole numbers and decimals.[30] Math may be a universal language, but it has many accents. American teachers with bilingual students in their classes need to be aware of these differences and the extent to which they may serve as subtle stumbling blocks in their math education.

One investigator found a distinct advantage for students who learned math in a single language. He compared two groups of Chinese-speaking 12-year-old students in Hong Kong who all attended Chinese schools through elementary school, had comparable test scores across subject areas, and took the same secondary school entrance examination. By parental choice, some students proceeded to Chinese high schools, while others moved on to more rigorous Anglo-Chinese high schools, where instructors taught in English. The two groups of students then spent the next 5 years studying in different languages; at the end of high school, they all took the Certificate of Education Examination. Those students who had attended the English-language high schools scored higher than those from the Chinese-speaking schools in all subjects except math and Chinese. The teachers attributed the Chinese-speaking students' math success in part to their language's superiority in expressing abstract math concepts.[31] Although researchers did not compare these two groups to monolingual English-speakers, one may speculate (alternatively or additionally) that the students who switched classroom language midway through their education may have found it difficult to link the math concepts learned earlier in one language with those learned later in another.

The work to date has been sparse, but these preliminary findings would suggest that, in the long run, mathematics education may be most effectively conducted entirely in the anticipated language of adulthood. In cases where immigration precludes this approach, students may need to relearn some elementary material in their new language. More research on this issue is urgently needed.

Left to Right, or Right to Left?

Chapter 2 discussed how our brains encode numbers along a mental number line. If numbers are organized spatially and sequentially, does that mean they increase in a particular direction? The answer, which science has not completely resolved, may be partly associated with reading and writing. This section also examines potential sources of confusion in the directionality of mathematical notation.

If a society's tape measures and rulers are any indication, Westerners expect to find zero all the way over on the left and big numbers to the right—an observation confirmed in the laboratory.[32] One tends to think of many sequences in a linear array, including the alphabet and the months of the year; as with numbers, Westerners "look for" A and *January* on the far left and Z and *December* to the right.[33] Researchers have speculated that people think of numbers growing in this direction in part because of the left-to-right visual scanning habits developed for reading and writing. A study of American preschoolers, however, suggests that children pick up the left-to-right habit even before they can read and write. When researchers showed a group of preschoolers a row of poker chips and asked them to add a chip or take one away, even the left-handed children did so at the rightmost end of the row.[34] Other observers have found that most young children begin their finger counts on the left hand.[35] These findings suggest that teachers' and parents' directional habits or some other related factor may have already influenced these young children's mental image of number.

What about cultures in which the language reads right to left? Clever experiments have shown that readers of English and French think of numbers growing from left to right, while readers of Arabic and Farsi, languages that scan from right to left, imagine numbers increasing in the opposite direction. Furthermore, Arabic and Farsi readers who later learn to read a Western language find that their clear right-to-left image of numbers diminishes as their Western reading becomes more fluent.[36] Deaf students also present an interesting situation: They read text and execute sign language from left to right but receive sign language from right to left. Although studies show that they have a left-to-right concept of number, they compare values more slowly than same-age individuals who are not deaf.[37] One wonders whether directional confusion contributes to their lack of fluency.

Learning conventional mathematical notation places a significant burden on children's directional orientation. Figure 7.1 illustrates the conflicting directions in which people read and write numerals on the page, as well as comprehend their values. For example, people read and write multidigit integers from left to right, even though their place values increase from right to left. Decimals produce more potential confusion because, although the place values diminish from left to right, the units (in English, at least) are referred to by larger and larger sounding number names (e.g., hundredths, thousandths)—a pattern consistent with the mental number line but the reverse pattern of that found on the whole-number side of the decimal point. Students new to decimal fractions often think that more digits imply bigger values—for example, *3.478 > 3.96*—as they do in whole numbers, or because hundredths are bigger than thousandths, that *3.47 > 3.927*.[38] Negative numerals, like positive numerals, increase in real value from left to right on the written number line, but their absolute values diminish as they approach zero from the left. Meanwhile, of course, the actual values of the individual digits vary unsystematically.

To make matters even more confusing in this era of globalization and immigration, numerical reading and writing directional trends vary across languages. Arabic

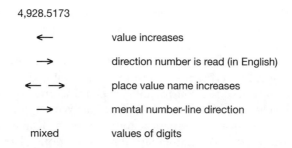

Figure 7.1. Directional variation in numerical concepts and notation, shown here for positive numbers.

language readers and writers, for example, who have the most consistent system, read *94* as "four plus ninety" and write it from right to left, with the *4* written first then the *9*, as they do in their alphabetic writing; they also write equations from right to left. (To clarify a terminology matter: "Arabic" numerals originated in India and got their name because they were brought to the West by the Arabs. Writers of Arabic actually use different numerals, although in the previous example, they are illustrated in the familiar form.) Iranians read and write numerals from left to right (opposite to their reading direction), as do Israelis, who incorporate left-to-right numerical statements into right-to-left text, disrupting the flow of both reading and writing. The Japanese use Arabic notation for most mathematical work but Kanji numerical symbols when they refer to quantities in vertical text. In South Korea, where Arabic notation is now the norm, discounts are advertised as "70%–30% sale," hoping to attract shoppers' attention to the first-read larger numeral on the left.[39]

What is known about the mental number organization of people from vertical-writing cultures? One study of Japanese students, who read and write both vertically and horizontally, found that their mental numbers arrayed themselves along both horizontal and vertical dimensions. Although their horizontal mental number line ran in the predicted left-to-right direction, in accordance with their horizontal writing, their vertical mental number line featured the small numbers at the bottom, like elevator buttons, rather than at the top where their vertical writing begins. Although some researchers think that this bottom-up vertical ordering of number may come from graphing convention, it is difficult to say which is the cause and which is the effect.[40]

One may speculate that general cultural or physical experiences at times trump reading and writing in determining sequential direction. For example, the numerals that come after *6* are to the right on a ruler and to the left on an analog clock.[41] And when given the choice between near and far, people think of small numbers as near.[42] Perhaps because gravity ensures that most things are anchored close to earth, people "see" small numbers at the bottom and big ones on top, like elevator buttons. Indeed, people commonly refer to "high" numbers and "low" numbers, certainly not "right" or "left" numbers.

The idea that experience with the physical environment naturally gives rise, so to speak, to a vertical sense of number has prompted some instructors to teach numbers vertically rather than horizontally. Thus one can easily evoke analogies to prices and temperatures rising and falling or to piles growing and shrinking and—when negative numbers are included—to values being figuratively above and below sea level. For children who need supplementary tactile and kinesthetic input, teachers have created these number lines from gift wrap tubing, enabling their students to slide their hands up and down the values. Coloring the negative numbers red may help children understand what being "in the red" means. Although as yet scientifically untested, these techniques may be especially useful for students with reading disabilities or certain ocular motor problems, for whom left-to-right scanning does not come easily. The efficacy of these methods—all logical extensions of number-line research reviewed previously—merits scientific investigation.

The most salient question, however, is this: What relevance does a directional number sense have for mathematical ability? Does a clear sense of direction on the mental number line make mathematical manipulations easier or more efficient, or does it matter? What about bilingual children who learn to read opposite-direction languages: Does number confuse them, or do their flexible scanning habits allow them a more flexible view of quantity? Researchers have additionally found that illiterate adults who can read only numerals do not demonstrate any directional bias for number or other sequences.[43] What is their idea of quantity, and what implication would that have for number-line development in individuals with severe reading impairments? How do children with impairments in scanning eye movement, visual sequencing, or left-to-right orientation understand numbers, compare numerical values, or master multidigit numerals and equations? Research on these compelling questions is in a very preliminary stage.

Fractions

As most middle school teachers can attest, fractions can bedevil even the best math students. Previously, we examined children's informal, intuitive grasp of proportion. This section discusses the formal aspects of fraction notation and terminology and explores some reasons why fractions are so troubling. The problem of fractions is addressed separately from other quantitative topics because they involve a fundamental shift in the way children think about numbers.

Upside-Down Numbers

The hardest aspects of learning fractions for many students are the spoken and written conventions used to describe and manipulate them. This is true not just for children with learning disabilities; the problem is widespread and persistent. Significant numbers of middle school students cannot rank order decimals or fractions, and some cannot even read them.[44] In one national multiple-choice math assessment, for instance, only 24% of 13-year-old students and 37% of 17-year-old students estimated $^{12}\!/_{13} + \frac{7}{8}$ to be about 2, whereas more than half of each group were either unsure or thought the answer could be 19 or 21.[45] These latter erroneous choices—the sums of the numerators and denominators, respectively—suggest these bewildered students were grasping at "buggy" fraction algorithms and had abandoned their fraction sense altogether.

Children's fraction difficulties begin with measurement and counting.[46] Students' earliest experience with proportion and conventional fractions is with continuous dimensions, or "stuff": string, pizza, sugar, and so forth. (Proportions of countable things, such as candies in a bag, come later, in part because children have other ways to apportion them.) Children are expected to use numbers to describe these proportions, but what do numbers have to do with measurement and proportion? How does one apply a tool developed for counting things to measuring dimensions?

The answer, of course, is to chop the dimension up into countable parts—slices of pizza, cups of sugar, miles of road, hours of the day. But other questions remain. For example, what size parts should be used? Continuous dimensions have no natural divisions (and, in fact, are infinitely divisible), so the size of the parts is necessarily arbitrary. Do the parts all have to be the same size? One central counting principle holds that objects' irrelevant features such as size do not matter: One Chihuahua plus one wolfhound equals two dogs. When numbers are used to measure, however, the parts must be the same size along the dimension being considered. The idea that one must use units to measure emerges around age 5 years, but it takes several years of formal schooling for young children to master the idea that they must all be the same size.[47] Students' potential confusion does not end once they understand that units must be the same size, however. When unit size remains constant, the proportional number of units may vary, as illustrated by the variable meaning of *one quarter* when speaking of dollars and hours in terms of cents and minutes.

Indeed, one reason that conventional fractions are so baffling is that they seem to turn the very meaning of number upside down and throw out the counting principles. In addition to the unit-size issue, for example, fractions do not have a natural counting sequence, nor do they have a unique number for each quantity (e.g., $\frac{2}{6} = \frac{1}{3}$). Comprehending a fraction (e.g., $\frac{3}{7}$, or three of seven parts) also involves double counting, another rule violation. (In this sense, part–part ratios—3:7—a simpler application of counting, are less confusing and hence are often grasped before fractions.) Moreover, a big number signifies a small amount ($\frac{1}{8}$) and a small number means a big amount ($\frac{1}{2}$)—a contradiction not only of the counting principles but of number sense! To cope with this latter problem, people often just ignore some of the values. This was probably what many of the students did when they misestimated $\frac{12}{13} + \frac{7}{8}$ in the previous example. It was also how a group of adults likely reasoned when researchers asked them to mark on a line how much a person earning $1 per hour (60 minutes) makes in relation to the amount made by a person earning $1 per minute or another earning $1 per day (1,440 minutes), as shown in Figure 7.2. Mathematically speaking, the relationship between the value of the denominator and that of the fraction is a power function, placing $\frac{1}{60}$ much closer to $\frac{1}{1,440}$ than to 1. Instead, the study participants responded as if they were comparing only the denominators (the number of minutes), placing $\frac{1}{60}$ closer to 1 as if the function were linear like integers. The correct answer is that $1 per hour is very close to $1 per day—a worker would take a job at $1 per minute in a heartbeat over a job at either of the other rates. The participants performed much better when the same question was put to them in familiar, linear decimal form in terms of dollars per minute: How does $0.0167 compare to $1.00 and $0.0007?[48]

Certain transformations are similarly paradoxial. Multiplying a counting number produces a bigger number (e.g., $6 \times 2 = 12$) and dividing results in a smaller one ($6 \div 2 = 3$); when it comes to proper fractions, however, multiplying produces a smaller number ($6 \times \frac{1}{2} = 3$) and dividing results in a bigger one ($6 \div \frac{1}{2} = 12$). The same principle, of course, applies to decimal fractions. Researchers asked a group of middle schoolers to solve the following problem: "A cheese weighs _____ pounds. One pound costs _____ dollars. Find out the price of the cheese." When the weight and cost were whole numbers, the children readily chose to multiply. But when the amounts were decimal fractions, fewer than one third of the students chose multiplication, opting instead to divide or subtract, even when the interviewer alerted the children to the similarity between the two versions of the problem.[49] Even some college students have trouble with the idea of a fraction as a multiplier. As with word problems and calculations in general, middle schoolers and high

Figure 7.2.　Money line. If the amount of money a person makes at $1 per minute is at one end, and the amount a person makes at $1 per day (1,440 minutes) is at the other, where would be the amount of money earned by someone making $1 per hour (60 minutes)? (Reprinted from *Journal of Experimental Child Psychology*, Vol. 101. C.A. Thompson and J.E. Opfer. Costs and benefits of representational change: Effects of context on age and sex differences in symbolic magnitude estimation, page 29. Copyright 2008, with permission from Elsevier.)

schoolers have even more difficulty with fraction word problems than with fractional computations.[50]

Of all the fractional operations, dividing by a fraction (e.g., $2\frac{1}{4} \div \frac{3}{4}$) is by far the gnarliest. The algorithm—multiply by the reciprocal ($2\frac{1}{4} \times \frac{4}{3}$)—is relatively easy to remember and execute. But what does it mean? The difficulty with the transformation is that it models at least three different kinds of seemingly unrelated problems. First, it solves measurement problems of the sort: "If a track is $\frac{3}{4}$ of a mile long, how many laps do I have to run to go $2\frac{1}{4}$ miles?" If the dividend were smaller than the divisor ($\frac{3}{4} \div 2\frac{1}{4}$), however, the problem would take the form, "What part of my $2\frac{1}{4}$-mile run is one $\frac{3}{4}$-mile lap?" Second, it solves this sort of problem: "If $\frac{3}{4}$ of a race is $2\frac{1}{4}$ miles, how long is the race?" This is known as the partitive model. Integer division finds the value of one part when the value of several parts is known (e.g., "If the length of a four-lap race is 3 miles, how long is each lap?"); in the fraction version, it finds the total value when the value of one part is known. Third, it solves problems of the sort: "If I run around a rectangular park that is $2\frac{1}{4}$ square miles, and two sides of the park are each $\frac{3}{4}$ of a mile, how long are the other two sides of the park?" This is the product-and-factors model of fraction division.[51] A student might struggle with any one of these concepts and thus could be forgiven for failing to intuit that the same operation can solve them all.

Not only are fractional notation and arithmetic difficult, but fraction-word grammar and vocabulary contribute to the confusion as well. For example, even though a numerator greater than 1 indicates multiple fractional parts, as in *three sevenths*, English speakers generally use a singular verb: *Three sevenths is bigger than two sevenths.*[52] As with number words, East Asian fraction terms are mathematically more transparent than Western ones. For example, in Japanese, *three quarters* translates as *of four parts, three.* In this way, one starts with the total number of parts (identified as such), giving instant meaning to the second term. In Korean, the expression translates more specifically as *of four equal parts, three*—the additional word being a common household term for *same size portions.* This distinction is significant: In a small study of sixth graders—two monolingual English-speakers and two bilingual children born in Korea and educated in the United States—the Korean children understood the importance of equal-size portions, whereas the American children focused entirely on the number of pieces in their fraction drawings.[53] Additionally, East Asian fraction number words are the same as those used in counting, avoiding the confusion found in English with words such as *half* or *quarter* or with a plural ordinal (e.g., *sixths*) in the denominator. In fact, cross-cultural studies of young students confirm that East Asian students learn conventional fractions much earlier than Western students. Researchers asked a group of American students in grades 4–8 to explain why fractions have two numbers; they found that only the gifted children could do so. When the English fraction terms were reworded, however, the American children performed as well as Asian students.[54]

Learning fractions is like stepping into the upside-down world beyond Alice's looking glass. No wonder children are confused!

Fraction confusion +

Chapter 3 introduced 14-year-old Cassie and described her visual-spatial difficulties and her superior iconic visualization. Here we address her frustration with fractions.

Most of Cassie's current math course has dealt with fractions and decimals, yet despite intensive instruction she does not understand their relative values. For example, when asked to order these decimals and fractions from least to greatest, she produced the following series: $0.2 \ldots {}^{45}/_{10} \ldots 0.12 \ldots 0.33 \ldots 0.127$.

According to her teacher, Cassie sequences values correctly about 40% of the time. When she attempts to compare fractions using diagrams, she fails to use equal-size wholes, making visual comparison more difficult, as in the following example.

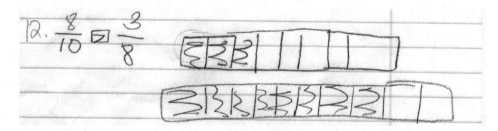

Cassie's incomprehension of fractional values is evident in where she locates fractions on a 0-to-1 number line. She was asked to place a randomly ordered series of fractions on individual number lines. One can see her confusion in the following composite diagram, where only ½, ¹⁄₁₃, and ⅞ are in or near their correct positions.

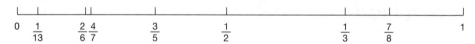

Cassie's difficulties raise the interesting question of whether one's general spatial skills, such as mental rotation, are linked to the quality of one's mental number line. (As previously noted, the two functions share a region of the brain.) Even Cassie's whole-number placements demonstrate confusion, as illustrated in the composite of her 0–1,000 number line placements. Unfortunately, research has not yet answered that question.

So far, the research on explicit number line instruction to enhance number skills has been limited to young children and whole numbers. It remains to be seen whether intensive teaching about the 0-to-1 number line would help an older student—particularly one like Cassie who has significant spatial impairments—clarify her understanding of fractions.

Teaching Fractions

As most intermediate grade and middle school instructors will attest, teaching fractions can be a frustrating and thankless task with disappointing results. Moreover, little rigorous research is available to guide pedagogy.

Notably, Chinese students have been particularly successful in learning fractions, a fact that some observers have attributed to their curriculum. According to a report from the 1990s, the Chinese national curriculum introduces fractions in the fourth grade and teaches the topics in a fixed order. First, they teach the concept of a fraction and then introduce decimals as special fractions with denominators of 10 and powers of 10. Once students thoroughly understand those concepts, they learn the four basic operations with decimals because of their similarity to whole-number operations. Teachers then introduce whole-number topics related to fractions, including divisors, multiples, prime numbers, prime factors, greatest common divisors, and lowest common multiples. Mastering those skills allows students to then learn such concepts as proper and improper fractions, mixed numbers, reducing fractions, and finding common denominators. Only when students have conquered those topics do they proceed to arithmetic operations with fractions.[55] Just as a thorough knowledge of whole numbers is crucial to arithmetic success, a mastery of fractions is prerequisite to learning fractional arithmetic.

The most difficult and important task is linking a written fraction to a useful mental representation of its value. Like whole numbers, fractions can be represented along a number line. In fact, researchers have met with success teaching fifth and sixth graders decimal fractions using a 0-to-1 number line. In this trial, two important instructional components included computerized feedback and requiring students to explain why their placements are correct.[56] Fraction blocks, comparable to whole-number blocks, are also available in common denominations. Unlike fraction "pizzas" and rectangles, they illustrate fractions as linear proportions of a given length, akin to a 0-to-1 number line. Thus they can be placed side by side and compared, fostering a sense of relative fractional value. To our knowledge, their pedagogical efficacy has not yet been tested, but should be. It will also be essential to determine for which students these approaches are most helpful.

Conclusion

To return to an earlier analogy, for some lucky people the language of mathematics is as eloquent as poetry; for many others, however, it is as baffling and infuriating as the U.S. tax code.[57] Conventional mathematics, including both terms and symbolic notation, allows one to move from rough, approximate reckoning to precise and sophisticated enumeration, calculation, and beyond; in that sense, it is a boon to mathematical thinking and creativity. Unfortunately, many students instead experience these formal conventions as disconnected from intuitive concepts and thus as an impediment to understanding.

Some students, including many with developmental language disorders, struggle to learn the counting sequence or find the technical language and phrasing poorly aligned with mathematical concepts. In addition to children with developmental language disorders, bilingual students and those from impoverished linguistic backgrounds are often particularly confused by this difficult and frequently inconsistent tongue. Because some math knowledge is intrinsically wed to the language in which it was learned, many bilingual students struggle to integrate mathematical concepts learned in different languages.

Likewise, while some individuals find conventional mathematical notation elegantly concise, others get lost in its forest of numerals and operational and relational symbols or become disoriented when asked to read, write, and think of numbers in different directions. Fractional notation, which they experience as only tenuously related to their intuitive sense of proportion and which often defies both whole-number sense and the counting principles, is particularly vexing for many students. These difficulties raise an interesting question regarding the relationship between math and reading. The two skills share some underlying cognitive and perceptual skills; thus, math disability and dyslexia often co-occur. Reading and writing alphabetic text and numeric text also rely on independent mental abilities, some children have great difficulty with one format but not with the other.

The solution to these challenges depends on future neurocognitive investigations into the way the brain processes formal mathematical language, including the technical terms and the formal notation, and how those functions overlap or are independent from natural language and reading. It also remains to be discovered whether methods found successful in teaching natural language and alphabetic reading to children with language disorders and dyslexia can be adapted to enhance mathematical language and reading skills for those who struggle with mathematical conventions.

Notes

1. Carroll, 1872/1960a, p. 129.
2. Henry, 2005.
3. Miles, Haslum, & Wheeler, 2001.
4. See Noonan, 1990.
5. See Durkin, Shire, Riem, Crowther, & Rutter, 1986.
6. Semple, 1992.
7. Stern & Stern, 1971.
8. Warren, 2003, p. 27.
9. Pimm, 1987, p. 90.
10. Thanks to Karen Haylor for this.
11. Stern & Stern, 1971.
12. Orr, 1987.
13. Perera, 1980.
14. See Kane, 1968; Munro, 1979.
15. Miller, Kelly, & Zhou, 2005.
16. Henry, 2005.
17. Stern & Stern, 1971.
18. Vander Linde, 1964. See also Rothman & Cohen, 1989.
19. Moats & Farrell, 2005.
20. Abedi, Hofstetter, Baker, & Lord, 2001; Abedi & Lord, 2001.
21. Hudson, 1983.
22. Warren, 2003, pp. 27, 29.
23. Jones, 1982; Lewis, 1989.
24. Jones, 1982; see also Spanos, Rhodes, Dale, & Crandall, 1988.
25. Fuson & Willis, 1989; Lewis, 1989; Zawaiza & Gerber, 1993.

[26]Spelke & Tsivkin, 2001; see also, e.g., French-Mestre & Vaid, 1993; Ginsburg, Posner, & Russell, 1981.

[27]Steingold, Spelke, & Kittredge, 2003.

[28]Dehaene, 1999.

[29]Lim, 2001.

[30]Spanos et al., 1988; M. Burton, personal communication, December 5, 2006.

[31]Chan, 1981.

[32]Dehaene, Bossini, & Giraux, 1993; Gevers & Lammertyn, 2005; Hubbard, Piazza, Pinel, & Dehaene, 2005.

[33]Gevers, Reynvoet, & Fias, 2003.

[34]Opfer & Thompson, 2006.

[35]Conant, 1896/1960, as cited in Fias & Fischer, 2005.

[36]Dehaene et al., 1993; Zebian, 2005.

[37]Bull, Marschark, & Blatto-Vallee, 2005; Iversen, Nuerk, & Willmes, 2004.

[38]Resnick et al., 1989.

[39]Zebian, 2005; M. Burton, S. Gurari, E. Gurari, N. Kayne, S. Kwon, and P. Tafti, personal communications, 2006–2007.

[40]Ito & Hatta, 2004.

[41]Bächtold, Baumuller, & Brugger, 1998.

[42]Santens & Gevers, 2008.

[43]Zebian, 2005.

[44]Mazzocco & Devlin, 2008.

[45]Carpenter, Corbitt, Kepner, Lindquist, & Reys, 1981.

[46]See Mix, Huttenlocher, & Levine, 2002b, the source for the otherwise unattributed material in this discussion.

[47]Petitto, 1990.

[48]Opfer, Thompson, & DeVries, 2007.

[49]Ekenstam & Greger, 1983.

[50]Carpenter, Corbitt, Kepner, Lindquist, & Reys, 1980, pp. 31–39; Hardiman & Mestre, 1989.

[51]Ma, 1999, pp. 55–83.

[52]Pimm, 1987, p. 83.

[53]Lim, 2001.

[54]Miller et al., 2005; Miura, Okamoto, Vlahovic-Stetic, Kim, & Han, 1999; Paik & Mix, 2003.

[55]Ma, 1999.

[56]Rittle-Johnson, Siegler, & Alibali, 2001.

[57]Perera, 1980.

Section **III**

Solving Problems

So far we have discussed how the human mind registers quantity and space and communicates mathematical ideas. The *doing* of mathematics, however, involves more than just number knowledge. It also requires the highly complex web of general-purpose mental abilities that one draws on for all intellectual activity and general problem solving.

Some of these skills are so basic that most people do not think about them. Psychologists call them *executive functions* because they keep one's mental life running efficiently and adaptively. These skills include focusing and sustaining attention, fending off distractions, controlling impulses and habits, and holding onto information. They also include planning ahead, monitoring oneself, and thinking flexibly. Psychologists liken these functions to the supervision and control exerted by a company's chief executive officer, an orchestra's conductor, or mission control for a space flight.

Solving mathematical problems also requires higher level reasoning—thinking abstractly and reasoning logically, both tasks that depend on solid executive skills. Taken together, these higher mental functions are related to what psychologists call "fluid intelligence," which, broadly speaking, has to do with the ability to solve problems. In particular, they have to do with *how* problems are solved, not *what* problems are solved, and thus apply to many settings, in school as well as in life.[1] As such, they are not specific to mathematical thought, but are necessary for it.

Unlike the number-specific abilities discussed in previous chapters, which are chiefly governed by the back two thirds of the brain, all of these general faculties involve the brain's frontal region, a portion of the brain that is much larger in human beings than in other animals. They also rely on vast neural networks connecting the frontal lobe to the brain sites that govern perception, cognition, emotion, motivation, movement, and memory.[2] Recent research has demonstrated frontal lobe activation during even the most basic mathematical tasks, as well as heightened activity there among math-talented students.[3] Emerging results of child studies have shown that the frontal lobes are among the last brain regions to reach maturity, typically in early adulthood.[4] Indeed, most of the skills governed by this area develop progressively throughout the preschool and elementary years, and many continue to mature during adolescence. Children develop these abilities at their own pace, and teachers will find considerable maturational variability within each classroom and across skills within each child.[5]

Chapter 8 discusses the mathematical and educational importance of the executive functions, the role of language in these functions, and some of the clinical con-

119

ditions that can impair them. Chapter 9 explores the higher-order thought processes most commonly required in mathematics.

No bright line separates many of these various mental activities. Although they are distinct, some depend on others and many overlap under certain circumstances.[6] For this reason, psychologists have had difficulty devising pure tests of any of them. Researchers have even had trouble coming up with a definition of executive function that everyone can agree on: Some psychologists emphasize self-control, others goal-directed behavior, and still others problem solving.[7] We focus our attention on the way these general mental activities support math learning.

Notes

[1]Mackintosh, 1998, pp. 297–330.

[2]See, e.g., Carpenter, Just, & Reichle, 2000; Gray, Chabris, & Braver, 2003.

[3]See, e.g., Dehaene, Cohen, & Changeux, 1998, pp. 246–251; Menon, Rivera, White, Glover, & Reiss, 2000; O'Boyle et al., 2005; Prabhakaran, Rypma, & Gabrieli, 2001.

[4]Bennett & Baird, 2006; Luciana, 2003.

[5]See, e.g., Barkley, 1997, pp. 218–227; Welsh, Pennington, & Groisser, 1991.

[6]Miyake et al., 2000.

[7]See, e.g., Barkley, 1997; Welsh et al., 1991; Zelazo, Carter, Reznick, & Frye, 1997.

Executive Functions

The mental abilities that people are most likely to take for granted are the ones that govern behavior. Math teachers rely on students to pay attention and remember what is said, to come to class prepared with pencils and completed assignments, to stay seated during class, and to check their work. When students do all these things, teachers can go about their business teaching math. When a student does not (or cannot) focus on the lesson, keep track of information and obligations, or think before responding, the student's work suffers and both student and teacher become frustrated. This chapter describes some executive functions and discusses their relevance to mathematics. It also explores the effect of impairments in these abilities on math learning and reviews some classroom techniques for reaching students with such impairments.

Attention and Mental Control

> Now I will have less distraction.
>
> Leonhard Euler, 18th century mathematician,
> upon losing the use of his right eye[1]

Any preschool, elementary, or middle-school teacher can attest that the ability to sit still, focus on one activity, screen out ambient and mental distractions, and persevere until the job is done emerges very slowly during childhood. Learning new information, such as $3 + 4 = 7$, relies upon the ability to focus on the given digits, operation, and result. It also requires screening out other digits, operations, and results, as well as the myriad compelling but math-irrelevant thoughts and distractions providing stiff competition for a child's limited attentional resources: the fly buzzing overhead, the child talking at the next desk, plans for recess, and so forth. In fact, a comprehensive study found that attention was the most robust factor affecting first graders' math performance.[2] What psychologists call *selective attention* entails the ability to focus on relevant information and to resist intruding distractions; *sustained attention*, or vigilance, is the ability to do this over time. *Distractibility* is the relative absence of these skills.

Not only must children screen out irrelevant information (i.e., control the input), they must also control their responses (the output). Math achievement, like all healthy adaptation, requires appropriate responses to events: correct answers, in the case of math problems. When students cannot screen out distracting information, they often have trouble not responding to it. It will not come as a surprise to any preschool teacher that one of the best predictors of early math success is self-control. Researchers gave a basic math test, as well as several nonmath tests requiring self-control, to healthy 2- to 5-year-old children. They found that math scores were directly related to impulse control, even when researchers took into account the child's age and vocabulary level, as well as the mother's education (a frequent predictor of school success).[3] Mental, as well as physical, self-control affect academic performance, and preschool teachers are not the only educators who need to be concerned about the effect of impulsiveness on math learning. Perseverance and self-regulation are significantly linked to academic success in general and math success in particular among students of all ages.[4]

Inattentive and impulsive children often seem to take the path of least resistance—miss an operational sign, add out of habit when they should subtract, or grab the first memorized fact that comes to mind. For example, researchers have shown that addition-multiplication fact substitutions (e.g., $7 \times 5 = 12$) decrease when children take more time to respond, suggesting that the extra time gives them a chance to inhibit the incorrect association.[5] Adults looking over the shoulders of these children as they work often feel compelled to interject, "Take your time," "Now pay attention," or "Think!" Teachers are good observers of their students' behavior. Several studies have demonstrated a link between students' arithmetic ability and teachers' ratings of their attention and persistence.[6]

Psychologists debate whether the brain's executive functions are independent of one another or are unified by some common feature. According to one philosophically appealing model, the primary executive function upon which all others depend is the ability to inhibit one's own behavior. Delaying one's response introduces into the mental universe a fourth dimension—time—and permits one to adapt to it.[7] In the math classroom, self-control buys the student time to think.

Working Memory

"Can you do Addition?" the White Queen asked. "What's one and one and one and one and one and one and one and one and one and one?"
"I don't know," said Alice, "I lost count."
"She can't do addition," the Red Queen interrupted.
From *Through the Looking-Glass*, by Lewis Carroll[8]

Working memory is the ability to hold in one's head bits of information, such as numbers or spatial configurations, while computing or reasoning with them.[9] Psychologists distinguish between *short-term memory* and *working memory*. Short-term memory entails simply holding on to information briefly in the face of distraction, as when a person looks up a telephone number (e.g., *614-555-1212*) and then must walk to the phone across the room to dial it. Working memory, on the other hand, entails holding the information in one's head as one manipulates it. Take as a silly but instructive example the requirement that, instead of just remembering the telephone number, one had to mentally reverse it and add 1 to each digit (*3232-666-527*). Such tasks require working memory and considerably more focus. Like poor Alice, one might "lose count."

Early models of memory distinguished between short-term memory, which involves temporary brain activation, and long-term memory, which entails brain cell growth over a period of time. Models now describe a more complex short-term working memory system composed of separate phonological and visual faculties (referred to as the *phonological loop* and the *visual-spatial sketchpad*). Brain and behavioral studies suggest that small bits of information can be stored for a few seconds in the back of the brain, close to where it perceives and integrates the verbal or visual input. To retain information past that time limit, however, a person must somehow refresh it. That work is done chiefly in the frontal regions of the brain, where cells can remain active for longer intervals.[10]

The front of the brain plays a crucial role in orchestrating these complex mental processes by focusing, dividing, sustaining, and switching attention as needed. Thus holding on to information relies heavily on attention and mental control; psychologists refer to this aspect of working memory as the *central executive*. In addition to directing attention, the central executive also supports organization. For example, breaking information into chunks (e.g., the "614," "555," and "1212" in that telephone number) makes it easier to remember and significantly increases the amount that people can keep in their heads. Through circuits linking the frontal lobes with long-term memory centers deep inside the brain, the central executive also directs the search for relevant long-term memories, both enhancing working memory and allowing the mind to create fresh ideas and connections. For example, after a few years of high school algebra, students can hold on to about twice as many symbols that are organized in meaningful strings as those arrayed nonsensically.[11]

Working memory requires considerable mental energy; the brain cannot inhibit interference or retain information indefinitely, and memories slip away. Math achievement depends on using this mental energy efficiently. Most mathematical activity requires working memory, although some problems require more of it than others. Tasks executed automatically require less attention and working memory than unfamiliar ones do, like the difference between relying on speed dial versus a telephone book. If less working memory is needed for one set of activities, then more is left over for others. Therefore, one argument in favor of mastering basic number skills is that the student can spend the salvaged attention and working memory on reasoning and higher-level problem solving. Students' basic number skills vary widely, even as late as college; failure to master the basic skills takes its toll on problem-solving speed and accuracy and on broader math achievement.[12]

Verbal Working Memory

The phonological loop holds and rehearses verbal material, as well as certain aspects of visual information, such as objects, pictures, geometric forms, letters, and words that one may remember by their names, descriptions, and other verbal associations. It is governed by the same brain region that controls speech; behavioral studies suggest that one remembers verbal information by literally rehearsing it over and over in one's head, as many people do when they try to retain a phone number. Brain studies have demonstrated more activation in this region during complex calculations requiring trading than during simple computations, presumably because a person must remember partial results while mentally jumping back and forth across place values.[13]

Verbal working memory, found to correlate with math performance at all ages, plays an important role in even simple math tasks.[14] For example, one function of finger counting among the youngest students, whose verbal working memory is not yet well established, is to help them keep track of their place in the count. Among older students, mentally calculating $57 + 4$, laid out as a horizontal sequence, relies on verbal working memory, probably because it is difficult to envision the algorithm. It also takes longer when presented as $4 + 57$, presumably because it requires more working memory to transpose the numerals.[15] Complex multiplication, even as tested on the computer using the numerical keyboard, also depends on phonological working memory, as one might expect from what is known about the verbal nature of multiplication fact retrieval.[16] In fact, any complex mental verbal task involving multiple steps or bits of information, such as computations or word problems, requires good verbal working memory, especially when visual cues are not available.[17] Studies have shown that performance on math word problems related directly to the children's ability to simply recall the problem, either before or after solving it.[18] This should not be surprising, however, because it is easier to remember something one understands than something one does not. It is not clear if efficient working memory is a prerequisite for mathematical understanding or a reflection of it; one suspects the relationship works both ways. Further research may help to answer this question.

Finally, one can remember something verbatim, such as the exact values in a problem, or retain the gist of it, such as whether one value is greater than another, corresponding to the two ways of thinking about numbers. These memories do not necessarily intersect, so remembering "five" and "seven" may not help a person remember that one number is bigger than the other. Reading works similarly: One can remember a sentence verbatim without thinking about what it means, and vice versa. Thus, it may be necessary to draw students' attention to the relationship between the values in a problem if they are too focused on remembering discrete facts.[19]

Visual Working Memory

Visual working memory also plays a crucial role in mathematical activity. The brain holds visual information in "sketchpad" fashion chiefly in the right hemisphere, with iconic and spatial material coded separately along their respective visual circuits.

The spatial circuit and the working-memory skills it supports develop slowly and continue to mature well into adolescence.[20] Spatial working memory impacts math-related learning very early, however. For example, beginning even in the first year of life, maturation of spatial working memory enhances infants' ability to track objects and predict object motion; by their second year, it supports the object-locating skills needed for the earliest geometric ideas.[21] In the toddler years, children's increased ability to create mental images and track sequential events, as seen in the development of symbolic play, enhances not only children's object locating, but their understanding of equivalence, their predictions of arithmetic outcomes, and their numerosity comparisons with ever bigger values.[22] Conversely, some young children understand more than they can demonstrate because their working memory lags behind.[23]

Students often rely on iconic and spatial imagery to enhance their mathematical performance; studies have demonstrated a link between math achievement and working memory for digits, visual sequences, and spatial locations.[24] For example, many people depend on a kind of "mental blackboard" to track the steps of complex mental calculations, particularly those that involve carrying and borrowing, and perform better when the numerals are visible than when they are not.[25] Moreover, they execute complex mental computations more quickly if the numerals are aligned vertically, allowing them to envision the algorithm.[26] Iconic working memory is also important for mental estimations of written calculations. For example, one study showed that people seem to prefer rounding down (247 → 240) rather than up (247 → 250) when estimating to avoid having to envision a new digit (in this case, 5) while they reckon in their heads.[27] In algebra, understanding improves iconic working memory and students' ability to identify and parse complex symbolic expressions.[28]

Spatial working memory is especially important for subtraction, even when executed on the computer using the numerical keyboard—a characteristic consistent with what is known about the number-line compatibility of that operation.[29] Mental rotation of two- and three-dimensional figures also requires a great deal of spatial working memory; those who lack it execute these tasks slowly and inefficiently or resort to nonspatial strategies.[30]

Some people use visual imagery to shore up their verbal working memory—rehearsing the information verbally while manipulating it visually, for instance. Verbal information is necessarily sequential, with one word following another; thus subsequent words often interfere with remembering the previous ones. By contrast, static visual information—a simple diagram, for instance, that can be swallowed whole—may be more memory friendly. Studies have shown that combining modalities strengthens working memory: One can remember more bits of visually descriptive verbal information than bits of verbal information that cannot be visualized.[31] "Seeing" the image can help one remember the words. This strategy can be especially useful in doing mental math, where transferring some mathematical information out of equations or verbal statements into mental images (sometimes expressed in diagrams or even gestures) can reduce the verbal memory burden.[32]

Working Memory Difficulties

Unfortunately, one's ability to hold on to information is time limited. Working memories are fragile and easily disrupted by talking, listening to irrelevant words or music, or engaging in competing visual tasks, for example.[33] (Not all students are adversely affected in this way, however, although the reason for this variability is unclear.[34]) If a person counts or computes slowly, it is more likely that his or her thoughts will decay—another good reason to have basic information stored securely in long-term memory. (Here again, the Chinese are advantaged: Their number words for 1 to 10 are each a short, phonologically simple syllable, allowing students to retain more of them in working memory.[35])

Poor working memory and associated compromised executive skills characterize many students with math disability.[36] In the early years, they count slowly, continue to rely on finger counting for basic computations, and fail to master basic arithmetic facts (or may even "learn" erroneous ones). These difficulties can lead to debilitating computational inefficiency and impaired learning later.[37] Moreover, many children with additional reading or general language disorders often have phonological or visual impairments that affect their working memory for number words or numerals.[38] Sometimes difficulties that appear to be due to unreliable working memory are instead related to (executive) trouble in screening out irrelevant information. Such is often the case when children respond with accurately remembered details that are in the question but are not relevant to the correct answer.[39]

Some information is especially difficult to hold in one's head. For example, numbers that rhyme or otherwise sound alike (*thirty* and *thirteen*, for instance) often blur in memory, as do digits that look similar. Working on one step of a two-step problem can interfere with remembering the other step. Also, excess verbiage in text can crowd out relevant information and make explanations hard to follow from the beginning of the sentence to the end.[40] Word problems in particular pose a significant challenge to verbal working memory: By the time students read the question at the end, many of them have forgotten the information at the beginning. Problems, such as the following one, that introduce a lot of numbers up front and save the question for last are especially difficult to remember, because there is no guiding question around which one can organize the incoming information:

> Charlie had a bag with 8 red marbles, 5 yellow marbles, and 7 green marbles. He picked out one marble without looking. What was his chance of picking out a yellow marble?

Very few students can solve this problem having heard or read it only once. Another particularly vexing type of elementary word problem places the missing information first, sometimes only implicitly:

> Joe won 3 marbles. Now he has 5 marbles. How many marbles did Joe have before he won?[41]

In this problem, one is given to understand without being told so explicitly until the end, that Joe had some marbles before he won the three new ones. The problem requires students to hold open a place in memory for this implicit, missing information. Teachers should be alert to both the working-memory demands of their math problems and the working-memory limitations of their students.

Planning and Sequencing

Tackling problems, in math and in life, usually involves a sequence of steps, beginning with realizing that there is a problem to be solved, then developing a mental understanding of its nature, devising a plan, and ending with a solution. These steps are not just mental; they also include speech and bodily movements, such as raising one's hand to be called on in class or picking up the pencil and writing out the problem. Although people sometimes grasp a crucial insight whole, often they must first analyze the problem, breaking it down into its various parts. How a person does that may depend on where he or she thinks the solution is headed; thus, it helps to keep the goal in mind at all times. In fact, sometimes in strategizing, the hardest part is being mindful of the goal, particularly if one does not fully grasp what the goal is.

To solve most problems, a person must consider the supporting arguments sequentially. In fact, failure to do so may lead to a flaw in the argument and the solution—hence many teachers' demands to "Show your work!" Like the other functions reviewed in this chapter, arguments are orchestrated in the front of the brain and rely heavily on attention, working memory for the relevant numbers and relationships, access to long-term memory of the relevant rules, and the ability to keep the goal in mind. As is the case with all supervisory mental activity, the problem's format and modality are largely irrelevant: Arithmetic computations, word problems, mental math, algebraic solutions, and geometry proofs—even changing a tire on a car—all require organizing the available information, devising a strategy, and taking the appropriate steps to reach a conclusion. That is not to say that the brain makes all its plans the same way. In one study, adult volunteers were told they would be given two numbers, but were asked to decide in advance whether they would add or subtract them. Functional brain images showed that the decision-making process activated a somewhat different region than the one activated during the execution of the operation itself. Moreover, they could detect subtle differences in the activation sites of the two operations in the planning stage.[42]

Impairment in planning and sequencing can contribute significantly to poor math performance in obvious ways: befuddlement about where to start, failure to follow algorithmic steps, confusion about the expected result, and circular thinking—all painfully familiar experiences to many math students. One neuropsychological test of planning, the Tower of London, requires individuals to devise a sequence of steps to rearrange colored balls on a series of pegs, under certain constraints, to match a given arrangement. When researchers gave this test to children and adolescents, those with math difficulties performed significantly worse than did those with either reading disabilities or no learning disorders.[43] Without a plan for solving a problem and sometimes without even a clear goal in mind, students often resort to their only alternative: trial and error—an inefficient primary strategy that consumes time unnecessarily and frequently leads to dead ends.

Self-Monitoring

Following a plan is adaptive as long as it leads to the goal. But when the strategy heads elsewhere, it needs to be reevaluated. Solving math problems, like adapting to real-world demands in general, requires the insight to know when one is off course. This sort of critical judgment is useful not only in evaluating one's solution to a problem, but also in assessing the reasonableness of one's chosen strategies for arriving at it.[44]

Critical judgment impairments are common among impulsive and inattentive individuals who do not take the time to reflect on their own behavior and its consequences or, in this case, their mathematical problem solving.[45] One example, previously discussed, is the student who used a misunderstood algorithm to arrive unquestioningly at $13 - 6 = 13$. Not all students who fail to recognize their own unreasonableness, however, do so for general lack of thoughtfulness or critical judgment. Students with poor number sense may also use unreasonable strategies or arrive at unreasonable answers; for them, however, it is because they do not understand the values. This is not an either/or situation, of course; some students' troubles may stem from several sources. Teachers need to know their students well to understand the exact nature of their difficulties.

Mental Flexibility

It is not sufficient to understand that one is headed down the wrong path. One must also be able to turn around and find another way out. In doing math, it is often necessary to revise one's thinking as new information becomes available, to approach a problem from a different angle, or to resist responding automatically according to an obsolete strategy or habit when such a response would be irrelevant to the solution. This tactic, known to psychologists as *shifting response set,* is just as important for math and life as having a plan in the first place. A person's ease in abandoning irrelevant strategies in favor of more adaptive ones is the mark of mental flexibility. From the earliest grades, math depends on flexible thinking for such tasks as choosing the most efficient strategy for doing simple sums, reckoning the same sum horizontally or vertically, or translating smoothly between objects, diagrams, and notation in doing arithmetic. In the classroom and the workplace, being able to jettison algorithms or standard operating procedures for workable short cuts can save valuable time and effort and lead to creative solutions for old problems.[46]

Learning, by definition, demands mental flexibility. Consider learning what the equals sign (=) means, for example, a task previously discussed in Chapter 7. Young children first understand that the equals sign is the signal to perform the operation on the left and put the answer on the right (e.g., $a + b + c = ?$). Later, however, they learn that the equals sign more broadly means *is the same as,* as in the statement $a + b + c = ? + c$. Research has shown that some children have no difficulty expanding their notion of the equals sign to include equivalence, whereas others remain stuck on the earlier model. Scholars have not yet accounted for these individual differences between children, but it does appear that for those who do have trouble adapting to the new concept, the greatest difficulty occurs after they have been introduced to the first skill, but before they have completely mastered it. When their command of the first skill is still fragile, new notions just confuse them.[47] Once they feel secure in traditional addition, they can more easily move on to addition equivalence. The same period of confusion is seen with many children as they make the transition from addition to multiplication, which makes similar demands on pattern recognition, memory, and mental flexibility.

Math achievement is affected by cognitive inflexibility.[48] Sometimes students simply fall into habits, but for students who cling rigidly to one fixed strategy, teachers should ask themselves why. Some students may simply not have been adequately taught another strategy. Other students, however, for any number of the reasons discussed in this book so far, may have trouble understanding an alternative strategy or may lack the working memory required to execute it. For these students, a careful evaluation is warranted to elucidate the reasons for their single-mindedness.

The Effect of Language

Language plays an important role in attention, self-control, working memory, and the other executive functions. Through it, a person can remember information, interject time between perception and reaction, test alternative strategies, and reflect on one's errors. Because language is necessarily sequential, it enhances organization, planning, and sequencing. This contribution is evident, for example, in one study of children with complete or partial hearing impairments and attendant language delays, who proved to be slower than children without those impairments in solving even nonverbal puzzles requiring planning and sequential steps.[49]

The sequencing of words (syntax) and the sequencing of actions are controlled by neighboring sites in the brain's left frontal lobe. The region that is activated during complex, goal-directed behavior stores and retrieves knowledge about the sequential aspects of action. It is concerned with natural cause and effect; as such, it operates relatively slowly with long-range objectives. By contrast, the area controlling the syntactic aspects of speech and language comprehension, located at the front end of the language circuit, concerns itself with the sequential structure of language. Because it must do its work during the brief time a sentence is heard or spoken, it operates more quickly.[50] Notably, solving math problems, whether in verbal form or in conventional symbolic notation, involves both modes of thought—that is, quickly comprehending the question's syntax and then slowly using that information to organize the steps for solving the problem.

Despite language's generally salutary effects on overall behavior, its syntax frequently presents a serious stumbling block to mathematical problem solving. Chapter 7 discussed one such example: comparative word problems. This section focuses on the problem-solving confusion created by the misalignment between natural and mathematical syntax—between the order of the words and the sequence of the action. Translating verbal statements into mathematical ones can be especially vexing beyond just the students' understanding of the words and phrases. For example, *take two from eight* is properly written as $8 - 2$. When researchers asked students to write as a mathematical statement, *Find a number such that seven less than the number is equal to twice the number minus twenty three*, they found that the students frequently expressed *seven less than the number* as $7 - N$, erroneously borrowing the verbal syntax for notational (and solution strategy) purposes. Similarly, in a separate problem, many students incorrectly wrote *a is five less than b* as $a = 5 - b$, exactly as directed![51] In Chapter 9, we examine further how words can trap people in literal, concrete thinking.

Syntax affects comprehension in both natural language and math. The less complex the syntax, the easier it is to follow directions—something most attentive parents and teachers know intuitively.[52] The situation is no different for math word problems, for which even many college students prefer the information to follow a certain order, so that they can more easily understand it and translate it into symbolic form.[53] Students tend to perform somewhat worse on arithmetic word problems than they do on the same computations presented in conventional symbols, suggesting that natural language syntax provides a significant impediment to problem solving.[54] That is not to say that the syntax of conventional mathematical notation is necessarily easy. The density of symbolic language makes each syntactic cue critical: $x - y$ is not the same as $y - x$, and $3x$ means something different from x^3.[55] Natural language statements often have helpful, embedded cues, such that even the most distracted or disorganized student will get the gist of it. For example, one easily grasps *The lion that chased the giraffe had a big mane,* in part because one knows from life experience that lions chase giraffes and that manes belong to lions. Notational math statements lack these familiar cues; word problems often have them, but as Chapter 9 points out, they sometimes add to the confusion.

A matter of syntax +

Logan is a left-handed 16-year-old who was brought to our attention when his teacher became concerned that he did not fully understand how the number line worked. For example, he seemed baffled when she asked him, "What number when added to negative fifteen gives a sum greater than zero?" The teacher figured that a cursory glance at the number line should yield the answer, but it did not.

The teacher then reworded the problem, "What do you have to add to negative fifteen to get a number greater than zero?", eliminating the passive voice and making the student, not the number, the actor in the question. This time, Logan answered easily. Logan's difficulty was not with the number line but with the question's syntax. In retrospect, his teacher realized that most of Logan's difficulties in her class stemmed from either not understanding what she said or his own trouble in explaining his thinking.

Logan's parents became concerned about him when his language did not develop typically by age 2. Comprehensive language assessments in both second and eighth grade confirmed significant impairment in receptive and expressive language. By the time we got to know Logan in ninth grade, he was silent unless spoken to and then responded by mumbling. Although he knows the meanings of words in isolation, he struggles to put them together syntactically. For example, he cannot repeat even simple sentences verbatim—a test of syntactic competence and sophistication—although he often understands them well enough to paraphrase them. His verbal memory is poor; remembering even a short list of numbers in the correct order is difficult for him. Logan can follow directions, however, if he can use picture cues to guide and remind him.

In stark contrast to his impaired verbal skills, Logan has superb visual and spatial skills. His performance on a test of mental rotation measured above the 90th percentile for his age. Although he cannot remember sentences, he has no trouble remembering pictures. His number-line placements of both positive and negative numbers are accurate, and he can easily determine which of two numbers is bigger or closer to another number. Logan's childhood pastimes included LEGOs, PlayStation, computer games, and building things. He had also developed a passion for constructing elaborate paper airplanes; when he becomes frustrated in class, he folds paper. The difficult part of learning to make the planes for him is following and remembering the written instructions; freed from the verbal directions, he has no trouble sequencing the folds.

Frustrated by his language difficulties both in the classroom and in social situations, Logan has disengaged from school and even from speech therapy. His family is exploring a technical vocational program, where he can be expected to thrive in the hands-on environment. In such a program, he will also get a new chance to discover how math can help him solve real, concrete problems.

What makes a word problem readable? Educators have studied readability for decades, producing grade-level guidelines for school texts in most subject areas. Math text is denser than that in most other fields, but readability issues are similar. On the whole, reading experts have found that children at any given age read and best understand the syntax that they themselves would produce but struggle with the formal sentence structures found in books. Among the easiest sentences to follow include simple sentences (*Mary has two marbles*) and coordinate clauses joined by *and* (*Mary has two marbles and Robert has three marbles*), as well as simple questions, exclamations, and imperatives. By contrast, the most difficult structures include passive voice (*Ten marbles are given to Robert* or *Three is added to one*), as well as dependent clauses, especially when they begin the sentence (*Using a diameter of five inches, draw a circle*) or, even more confusing, when they scramble the action sequence (*Before Robert took two of her marbles, Mary had three*). Some educators have suggested that because simple sentences tend to be short, textbook writers would do well just to stick with short sentences. Others, however, have argued that short sentences are not always easy to understand, nor are long sentences always difficult. Because math text as found in word problems or textbook explanations is naturally dense, sometimes comprehensibility benefits from slightly wordier, but relevant, elaboration.[56]

Psychologically speaking, one of the factors contributing to sentence incomprehensibility is its demand on working memory. Grammarians address the working memory issue in what they refer to as a word's "depth," which is the number of grammatical structures one has started, but not completed, at the time one speaks or writes the word. For example, when students read the word *using* in *Using legs of three and five inches, draw a rectangle,* they must keep in mind that what follows (*legs of three and five inches*) will need to be used. Because this is a dependent clause, however, they must also hold the thought while they read the independent clause (*draw a rectangle*), which, at long last, reveals the goal of the instruction. The deeper the words, the harder the sentence is to understand and remember. To remember sequential instructions and plan a strategy, one needs an easy way to track the steps and components. For this purpose, serial sentences or coordinate clauses work better, particularly if they are sequentially aligned with the solution strategy. Also for this reason, it is important to clarify referents (e.g., in the sentence *Robert had two marbles, and Mary had three*, it may be unclear to what the word *three* refers).[57]

Studies have shown that word problem syntax affects math performance, not only among children with working memory impairments, but also among students new to English, those of lower socioeconomic status, and those with low to average math ability.[58] Math problems that read like cake recipes, presenting each bit of information in the order one needs to use it, leave little room for error due to poor working memory or for confusion stemming from a mismatch between syntactic sequence and solution sequence. Recipe-like problems also make clear who is acting upon what. Unfortunately, in real life neither the problem nor the path to the solution are typically that transparent, and students' skills on syntactically simplified problems do not necessarily generalize to more realistically presented practical problems.[59]

As children mature, their ever-improving verbal working memory, as well as their planning and sequencing abilities, allow them to think about math problems in increasingly more complex ways.[60] To translate those expanding ideas into formal mathematics, however, requires syntactic sophistication. In the context of a rich verbal environment, reading, and formal education, children's ability to understand and produce increasingly complex syntax in reading and speech normally improves with age.[61] Under ideal conditions, students learn to move back and forth easily between the syntax of natural language and that of mathematical notation and solution planning. Unfortunately, not all students grow up under such favorable circumstances or can benefit from them if they do.

Conditions Affecting Executive Functions and Math Learning

Selective and sustained attention, working memory, and the other executive functions are fragile and often deteriorate under stress. Children who struggle with math for any of the reasons cited in earlier chapters will naturally have trouble focusing on or remembering the material and may seem highly distractible, just as anyone would if required to do something for hours on end that they did not understand. The same is true for children experiencing general psychological upset or the mentally disorganizing effects of neuropsychiatric illness; math troubles characterize a wide range of psychological, neuropsychiatric, and other medical conditions. Space does not permit an exhaustive review, but we encourage interested readers to consult the appropriate references for more information on the math difficulties associated with these disorders.[62] We focus here briefly on anxiety (particularly math anxiety) and then more extensively on ADHD.

One of the least understood conditions affecting math performance is anxiety. Any compelling negative emotion or preoccupation, regardless of the reason—including anxiety—easily disrupts attention, mental control, working memory, and the other executive functions. Students who attempt to learn a math lesson or take a test while they are worried, upset, or preoccupied are usually flying into the wind. One type of anxiety, which has been extensively documented, is math anxiety—anxiety triggered simply by the demand to understand or convey math concepts. Studies have demonstrated that people with math anxiety have diminished working memory on math (but not nonmath) tasks, such as complex calculations involving trading, and are slow and inaccurate on counting and computations, particularly under time pressure. At more advanced mathematical levels, they fail to learn and do poorly on math tests. Research has not yet sorted out the question of causality, however. Does the anxiety—which often begins around sixth grade and seems to be tied to social pressures, past classroom experiences, and self-image issues—cause poor performance, or is it a rational response to being confronted by one's mathematical incompetence?[63] Both possibilities are reasonable; anxiety and disability likely feed on each other and are difficult to tease apart in research. Perhaps future investigations will unravel this knot.

The best-documented developmental neuropsychiatric condition featuring executive function impairments is ADHD—a developmental neurobehavioral syndrome characterized by distractibility, disorganization, poorly sustained attention, and forgetfulness, and/or by hyperactivity, fidgetiness, loquacity, impatience, and impulsiveness. ADHD is present in up to about 7% of school-age children and 4%–5% of adults.[64] Brain studies of individuals with ADHD have revealed abnormal activation patterns, particularly in frontal and parietal regions, during tests of attention and self-control.[65]

Few math-disability investigators have addressed ADHD or even accounted for it in their study designs, but one systematic study of fifth-grade students with severe math disabilities found that about one in four merited a formal diagnosis of ADHD. Following those students through high school, the researchers found that those whose severe math difficulties persisted were more likely to have had attention problems throughout their schooling, although the most severe math difficulties were not solely attributable to ADHD.[66]

Looking at the problem from the other side, the prevalence of serious math difficulty among children and adolescents with ADHD has been estimated between 21% and 55%, depending on researchers' working definitions.[67] ADHD (particularly the inattentive type) typically takes its severest academic toll on math and spelling; of all subjects, math is the one in which students with ADHD are most likely to fall behind their class.[68]

What math troubles do students with ADHD have? Unlike the difficulties reviewed in previous chapters, ADHD seems to affect performance rather than understanding, at least in the short term. For example, they often count slowly and have more than the usual difficulty in comparing numbers close in value; number-line placement, however, tends to be unaffected. But because math learning is cumulative and ADHD, as a developmental disorder, affects children from the earliest years, early performance failures can seriously compromise understanding downstream. Many such students fail to master basic arithmetic facts because they lack the focus or patience for the repetitive drill and lack the organization necessary to develop or follow a retrieval strategy. Not having elementary facts at hand puts an even greater strain on their already compromised attention and working memory in solving complex problems, particularly those with multiple operations or trading. Typically, students with ADHD compute slowly, in part because they have failed to master the basic facts, in part because their minds wander, or even because they are out of their seats doing unrelated things; on time-limited tests, the slow pace results in fewer attempted items. Complex problems often confuse these students because they have trouble keeping in mind the original question while organizing the information and planning the solution, and they have difficulty shifting strategies when necessary. Finally, students with ADHD have trouble both ignoring extraneous information and controlling their impulsive response style, leading to "careless" errors.[69]

One study put a small group of 7- to 12-year-old boys, with and without ADHD, under a behavioral microscope. The children, who had been referred to professionals for behavioral or attention problems, were of normal intelligence and had at least average-range arithmetic skills for their grade. Of those with ADHD, 25% had inattention symptoms only and the rest had a combination of inattention, hyperactivity, and impulsiveness; none were taking medication for their symptoms. The participants were given a set of addition and subtraction problems specially designed to be challenging yet within each student's ability; researchers did not time their performance, but gave them an ample 10 minutes to complete the test. The findings were instructive: The children without ADHD attempted twice as many problems as did those with ADHD and were also more accurate. The differences were especially pronounced on subtraction problems, on which those with ADHD made 6 times as many trading errors as did those without ADHD; in fact, most of those with ADHD consistently subtracted the smaller digit from the larger one, regardless of its position in the problem. Moreover, the students with ADHD were more fidgety, left their seats more often, were less attentive, used finger counting more often, and used pencils and other materials for nontest activity more frequently than did those without ADHD.[70]

A boy with attention-deficit/ hyperactivity disorder (ADHD) + + + + + + + + + + + + +

Taylor is a personable 12-year-old who loves video games and is interested in how electronics and cars work. He was brought to our attention because he is often confused in math class and has not yet mastered his math facts.

His mother's pregnancy was uneventful, but delivery was difficult. Taylor's gross motor skills were all delayed; currently, sports frustrate him because of his poor physical coordination. He was initially evaluated in the first grade because he talked out of turn and had trouble concentrating. His parents report that he has difficulty sustaining attention to chores, tests, and schoolwork. An evaluation at age 8 revealed very poor working memory and unreliable access to information held in long-term memory. At that time, he was diagnosed with ADHD. Taylor currently takes medication to control his inattention, but in class he still has trouble following directions and working independently, even when his teachers provide him with a written checklist. Recent assessment confirmed his trouble following even the most basic instructions.

Our own observations of Taylor during a brief evaluation were consistent with others' impressions. When asked to write the alphabet as quickly as he could, he stalled out at *V* because he lost his train of thought, rationalizing "I haven't written the alphabet since the first grade." He seemed to sense that he could not retain verbal instructions and asked to have every direction repeated; indeed, both his verbal and visual memory were poor. He copied nine geometric designs adequately, but could recall only three of them immediately afterward. Taylor took advantage of every few-second break between evaluation tasks to chat about his favorite video games, going to the movies with his dad, and whatever else came into his head, prolonging the interview significantly. Indeed, Taylor did not seem to find time particularly salient. When encouraged to take his time on one activity, he barely slowed down, yet when urgently reminded that he was being timed on another, he responded, "So?"

Taylor's extraordinary distractibility was evident on every math task. When asked to count by pointing to a row of objects, he lost track and double-counted the last one. He did not trust his own working memory and surreptitiously used his fingers to mentally subtract 3 serially from 35—and still lost focus, ending with ". . . five, four, three, zero." His long-term memory for mathematical information was just as scrambled. He successfully located $\frac{1}{2}$ on a 0-to-1 number line, but then when asked to indicate where $\frac{3}{4}$ would be, he marked the line near 0, breezily declaring that he knew where $\frac{3}{4}$ went because "all those numbers like that are less than a half."

While he was taking a timed test of simple addition, subtraction, and multiplication, a man in the hallway outside the room blew his nose—a long, loud series of honks. From that point on, Taylor slowed down; when the test was over, he declared, "That man blowing his nose was really distracting! I couldn't think! I pretty much know these facts, but I couldn't remember them at all and had to use my fingers!" He scored below the 10th percentile for his age on the test because he worked so inefficiently, and he missed 4 of the 58 problems he attempted, failing to notice the sign changes in 2 of them.

Taylor takes medication to help him focus, his teachers understand his need for structure and redirection, and when he can focus he picks up some math, which he usually understands. Not all students with ADHD are as fortunate.

A man with attention-deficit/ hyperactivity disorder (ADHD) + + + + + + + + + + + + +

An amiable young man, Spencer was 23 years old when he first saw a psychologist for an evaluation. At college matriculation, he had failed to place into the one math class required for graduation, meaning that he had to take two noncredit preparatory math classes before he could even sign up for the required course. By the time he was referred for evaluation, Spencer had failed the first noncredit class four times. He had already completed his other graduation requirements and was now afraid that he would never earn his degree.

Spencer's early development seemed fairly typical, but things were different when he started school and had to sit still and pay attention. He was frequently sent out to the hallway for being "rambunctious" and otherwise spent much class time staring out the window. He remained poorly focused, inattentive, distractible, forgetful, unable to follow through on tasks or instructions, and generally disorganized. At home after school, his parents tried eliminating distractions, such as the television and games, but would find him doodling instead of doing his homework and had to "get in his face" about doing his assignments.

Although a capable reader and inquisitive, Spencer rarely had the patience to pick up a book; when he did, he skipped around from paragraph to paragraph. Lacking sustained focus or the ability to hold on to information through the next distraction, learning number facts and algorithms was nearly impossible. (At age 23, Spencer still used his fingers for subtraction and required a calculator for multiplication facts, which he had never memorized.) His parents considered having him evaluated for ADHD when he was younger but feared medications, so they did not pursue outside help.

Undiagnosed and untreated, most of Spencer's problems persisted through adolescence. Without the fundamentals, the more abstract mathematics of middle school went over his head. In high school, sports always came before homework, as did myriad other internal and external distractions. Spencer required a remedial course and tutoring in ninth grade to pass the state proficiency test and failed geometry—his last math course before the string of college failures—with a grade of 18%. In college, he was fidgety and had trouble sitting through classes, getting up to walk around when he could not sit still any longer. Not surprisingly, his self-esteem suffered and was redeemed only by his athletic success. As college progressed, however, he found courses and a field he loved, and his grades improved—except in math.

The psychological evaluation at age 23 revealed a young man of average intelligence with severe, lifelong ADHD that had stunted his academic skills, most particularly math. During the examination, Spencer's attention waxed and waned markedly. At times he "zoned out" and could not remember anything for 5 seconds; at other times, his focus was intense, as is typical in this disorder. For the most part, however, his working memory was severely impaired and he missed important details. His poor attention and working memory, in turn, compromised his ability to think in logical steps and to reason effectively, leaving him with only concrete, trial-and-error strategies. His mastery of basic arithmetic measured in the 5th percentile—far below the level expected from his intelligence—and his difficulty in tracking multiple bits of information at once took a significant toll on his ability to think through applied problems or weigh solution options.

Spencer received the news of his ADHD diagnosis and average intelligence with relief, but he was also distressed when he thought of all the teachers who had told him he was lazy and stupid, and of all the years of education that he had lost. His college, not knowing how to help him catch up, eventually waived his math requirement, allowing him to graduate.

Classroom Implications

There is no cure for attention, working memory, or other executive impairments, but several approaches have demonstrated some improvement in classroom performance. These methods generally fall into two categories: medication for neuropsychiatric conditions and environmental alterations, including classroom strategies.

For ADHD, the medications most frequently prescribed work on a natural brain chemical.[71] Studies of elementary students with ADHD have shown that they used fact retrieval more often, attempted more test problems, and corrected themselves more reliably on medication than off. Researchers continue to debate whether medication improves the accuracy of the children's answers, however. Although some studies have found significant short-term performance improvements, others point out that medication did not help children recover lost ground in learning or change well-established error patterns. Moreover, gains noted when the children were on medication disappeared after its discontinuation.[72] Further research is needed to determine whether long-term medication use, instituted at the first sign of debilitating disruption, can help children master and retain early mathematical concepts and at what cost.

Nonpharmacological strategies are available to all students with executive impairments, whether or not they have an underlying neuropsychiatric condition. For math anxiety, researchers have tried various techniques (chiefly with college students), including cognitive behavioral therapy and having students overlearn the material. Unfortunately, the success rate across these different methods has been only about 50%, and even the techniques most successful in reducing anxiety have not improved students' math skills.[73]

For distractible children, computer fact drill can be effective, particularly when it includes immediate feedback, rewards, and occasional competition, although actual games are often too distracting. Judicious use of color and movement in computer graphics focuses the child's attention on the material and away from extraneous background. Curiously, background music has also been shown to help focus attention for children with ADHD, although the reason for these counterintuitive findings is unclear. Graphing one's progress and even videotaping one's own behavior enhance self-monitoring; standing up, moving around, and taking brief breaks help to discharge physical energy. For students with compromised working memory or sequencing skill, direct instruction may work more effectively than "discovery" methods for teaching new arithmetic procedures, because it reduces memory overload and organizes the information to be learned. Requiring students to accurately explain why answers to practice problems are correct or incorrect may help them focus and be better able to apply the procedure to unfamiliar problems. Distractible children have trouble with facts and procedures, but they can often understand the big ideas and should be provided with plenty of opportunity to explore those in small groups, verbally and through hands-on activity.[74] As with most pedagogical techniques we report, these all would benefit from further scientific testing.

One particularly successful method is "think-aloud" exercises, in which students verbalize each step of a solution. Training, which takes time, begins with the teacher's modeling the process and then guiding the children to do it for themselves. For children who are distractible, continuous verbalization focuses their attention on the problem. For those who are impulsive, it buys them time to think before they respond. For those with poor planning and sequencing skills, it structures their thought processes. And for most children, it allows them the freedom to explore various solutions. It is a method, once mastered, that children can apply to all sorts of problems—in life and in math.[75]

For students with working memory impairments, one tactic for increasing total working memory capacity is to distribute information across verbal and visual modalities. Reducing the amount of incoming verbal information by supplementing with diagrams and graphs, color coding steps in multistep problems, or even using calculators takes the load off limited verbal working memory. Some children, particularly those with reading disabilities who cannot remember what they read, are more likely to identify the correct arithmetic operation required by a story problem when it is presented as a drawing rather than in written verbal format. In one study of eighth graders, investigators found that teaching geometry proofs using a combination of diagrams and spoken explanation was more effective than either auditory or visual methods alone. Moreover, children learning in the combined modalities were then better able to apply what they had learned to new problems.[76]

Conclusion

The mental functions discussed in this chapter—attention, mental control, working memory, planning, self-monitoring, and mental flexibility—are fundamental not only to mathematical activity but to daily life. These executive abilities exert a powerful effect on math learning, but impairments in them are challenging for teachers to identify. Students may look like they are paying attention while their minds are elsewhere, and teachers often either rationalize other telltale behaviors—forgetting assignments, failing to master arithmetic facts or complete tests, making "careless" errors—or interpret them as signs of poor character. In fact, many distractible or disorganized students with compromised executive functions are highly motivated and expend more time and energy, not less, on seatwork than their classmates, although for fewer returns. When in doubt, teachers would do well to ensure that the child is professionally evaluated for conditions linked to executive function difficulties and to provide appropriate treatment and pedagogical adjustments.

Notes

1. Eves, 2003, 241°.
2. Fuchs et al., 2005.
3. Espy et al., 2004.
4. See, e.g., Bull & Scerif, 2001; Lindsay, Tomazic, Levinc, & Accardo, 2001.
5. Lemaire, Barrett, Fayol, & Abdi, 1994.
6. Deater-Deckard, Petrill, Thompson, & DeThorne, 2005; Fuchs, Fuchs, Compton, et al., 2006; Lundberg & Sterner, 2006.
7. Barkley, 1997; Miyake et al., 2000.
8. Carroll, 1872/1960b, p. 320.
9. See Adams & Hitch, 1998; Baddeley, 2003, for the reviews upon which this section is based.
10. See Grafman, 2002, pp. 159–160; Smith & Jonides, 1997.
11. Sweller & Cooper, 1985.
12. Ashcraft, Donley, Halas, & Vakali, 1992; Barrouillet & Lepine, 2005.
13. Gruber, Indefrey, Steinmetz, & Kleinschmidt, 2001; Smith & Jonides, 1997.
14. See, e.g., Bull & Scerif, 2001; Fuchs et al., 2005; Hitch, 1978; Lehto, 1995.
15. Trbovich & LeFevre, 2003.
16. Lee & Kang, 2002.
17. Fürst and Hitch, 2000; Hecht, Close, & Santisi, 2003.
18. Cummins, Kintsch, Reusser, & Weimer, 1988.
19. Brainerd & Gordon, 1994.
20. Klingberg, Forssberg, & Westerberg, 2002; Kwon, Reiss, & Menon, 2002.
21. Baillargeon, 1994; Newcombe, Huttenlocher, & Learmonth, 1999.
22. Mix, Huttenlocher, & Levine, 2002b; Sluzenski, Newcombe, & Satlow, 2004.
23. Fischer & Beckey, 1990; Klein & Bisanz, 2000.
24. Dark & Benbow, 1991; Kulp et al., 2004; Reuhkala, 2001.
25. Adams & Hitch, 1997.
26. Trbovich & LeFevre, 2003.
27. Lemaire & Lecacheur, 2002.
28. Sweller & Cooper, 1985.
29. Lee & Kang, 2002.
30. Carpenter & Just, 1986.
31. Brooks, 1968, as cited in Baddeley, 2003, p. 834.

[32]Goldin-Meadow, Nusbaum, Kelly, & Wagner, 2001.

[33]Adams & Hitch, 1998.

[34]Abikoff, Courtney, Szeibel, & Koplewicz, 1996.

[35]Hoosain & Salili, 1987; Stigler, Lee, & Stevenson, 1986.

[36]Passolunghi & Siegel, 2004; Wilson & Swanson, 2001.

[37]Geary, Hoard, Byrd-Craven, & DeSoto, 2004; Swanson & Beebe-Frankenberger, 2004.

[38]Jordan & Montani, 1997; Lachmann, 2002; Steeves, 1983; Ellis & Miles, 1977.

[39]Passolunghi & Siegel, 2004.

[40]Baddeley, 2003.

[41]DeCorte, Verschaffel, & DeWin, 1985, p. 464.

[42]Haynes et al., 2007.

[43]Sikora, Haley, Edwards, & Butler, 2002.

[44]See Siegler, 1988.

[45]See, e.g., Rourke, 1993.

[46]See, e.g., Scribner, 1986.

[47]McNeil, 2007.

[48]Bull & Scerif, 2001.

[49]Luckner & McNeill, 1994.

[50]Sirigu et al., 1998.

[51]Spanos, Rhodes, Dale, & Crandall, 1988, p. 222.

[52]Vander Will, 1976.

[53]Lewis & Mayer, 1987.

[54]Carpenter, Corbitt, Kepner, Lindquist, & Reys, 1980.

[55]See Munro, 1979.

[56]Noonan, 1990; Perera, 1980; Spanos et al., 1988.

[57]Bormuth, 1966; Noonan, 1990.

[58]Abedi & Lord, 2001. See also Wheeler & McNutt, 1983.

[59]Fuchs & Fuchs, 2002.

[60]See, e.g., Case & Griffin, 1990.

[61]Soifer, 2005.

[62]Acute lymphoblastic leukemia: Kaemingk, Carey, Moore, Herzer, & Hutter, 2004. Adolescent behavior disorders: Barriga et al., 2002. Bipolar disorder: Doyle et al., 2005; Lagace, Kutcher, & Robertson, 2003. Fetal alcohol syndrome: Kopera-Frye, Dehaene, & Streissguth, 1996. Low birth weight: Taylor, Burant, Holding, Klein, & Hack, 2002. Minor head injury: Levin et al., 1987. Schizophrenia and major depression: Hugdahl et al., 2004; Kiefer, Apel, & Weisbrod, 2002. Seizures: Seidenberg et al., 1986; Shalev, Gross-Tsur, & Masur, 1995. Tourette syndrome: Dykens et al., 1990.

[63]Ashcraft, Krause, & Hopko, 2007.

[64]American Psychiatric Association, 2000; Barkley, 1997, p. 3; Kessler et al., 2006.

[65]Pliszka et al., 2006; Smith, Taylor, Brammer, Toone, & Rubia, 2006; Tamm, Menon, & Reiss, 2006.

[66]Gross-Tsur, Manor, & Shalev, 1996; Shalev, Manor, & Gross-Tsur, 2005.

[67]Faraone et al., 1993; Semrud-Clikeman et al., 1992.

[68]Loney, Kramer, & Millich, 1979; Morgan, Hynd, Riccio, & Hall, 1996; Zentall, 1993.

[69]Biederman et al., 2006; Kaufmann & Nuerk, 2008; Marzocchi, Lucangeli, DeMeo, Fini, & Cornoldi, 2002; Zentall, 1993.

[70]Benedetto-Nasko & Tannock, 1999.

[71]Volkow et al., 2004.

[72]Benedetto-Nasko & Tannock, 1999; Douglas, Barr, O'Neill, & Britton, 1986.

[73]Beilock, Kulp, Holt, & Carr, 2004; Hutton & Levitt, 1987; Zettle, 2003.

[74]Abikoff et al., 1996; Christensen & Gerber, 1990; Rittle-Johnson, 2006; Zentall, 2007.

[75]Ross & Baden, 1991; Schunk & Cox, 1986.

[76]Mousavi, Low, & Sweller, 1995; Moyer, Sowder, Threadgill-Sowder, & Moyer, 1984; Zentall, 2007.

Chapter 9

Reasoning

> In order to solve this differential equation, you look at it until a solution occurs to you.
>
> George Pólya, mathematics professor,
> mocking the professorial stereotype[1]

Anyone who has struggled through a difficult math lesson knows that mathematics involves a lot more than staying alert. If that were the only requirement for success, there would be many more mathematicians. Proficient mathematical thinking involves what philosophers and psychologists call *higher-order thinking, fluid intelligence,* or *reasoning.* Although reasoning goes beyond simple attention and recall, it depends significantly on attention, working memory, and the other executive functions; like them, it matures slowly. (With effort, however, some individuals can mobilize powerful reasoning skills to compensate for unreliable supervisory control; for this reason, executive impairments are often missed in very bright students.[2]) Even the earliest learning, including language acquisition, depends on the ability to reason, which in turn is affected by experience. Influenced by myriad genetic and environmental factors, a person's reasoning faculties contribute not only to academic success but also to the globally adaptive capacities that place human beings at the fortunate end of the food chain.[3]

Reasoning is also the soul of mathematics. This chapter discusses two broadly defined higher-order mental skills—thinking abstractly and arguing logically—required for all mathematical problem solving.

Thinking Abstractly

> [N]umber applies itself to men, angels, actions, thoughts, every thing that either doth exist, or can be imagined.
>
> John Locke[4]

Mathematics is an abstract enterprise. Its business is to decipher and formalize into words and notation the patterns and rules that describe an orderly universe. These regularities are reflected in the counting principles, dimensional relationships (e.g., the trigonometric ratios, π), physical laws, even predictable patterns of human be-havior. Deciphering these regularities requires the ability to peel away irrelevant fea-tures to arrive at defining characteristics—as mathematician Jean Dieudonné noted, to engage in "the art of selective ignorance."[5] Number itself is abstract in that it refers only to the numerosity of objects or events, with their other traits stripped away; it thus can apply to any "thing that either doth exist, or can be imagined." Number words and numerals stand in for numerosities, not for those concrete ob-jects or events themselves. Categories are defined by abstractions: Whole numbers are examples of integers, for example, and rational numbers are instances of real numbers. Math problems can be categorized according to how one can represent them or the operation one might use to solve them.

Mathematics begins when objects are left behind and all that remains are their quantitative features. For example, consider the statement, *If I have two apples and buy three more, I will then have five apples.* Mathematically, the scenario can be repre-sented by the abstract statement *2 + 3 = 5*, in which the *2, 3,* and *5* stand for quan-tities of apples, not the apples themselves. In this sense, a mathematical statement is not concrete. Nor is it literal: The symbols themselves carry no meaning inde-pendent of the concepts they represent. Rather, mathematical words and symbols are "windows through which to see their referents," as educator David Pimm has noted.[6] Previous chapters illustrated how challenging it is to write a mathematical statement that means the same as the phrase *a is five less than b* or to escape the con-viction that $(x + y)^2 = x^2 + y^2$ just because it looks good. One can get similarly trapped by generic prototypes, such as the illustrations accompanying textbook geometry problems: Students sometimes assume that a triangle in a given problem is necessarily equilateral because the illustration in the textbook happens to look that way, for example. The textbook sketch is merely a visual aid to reason, not the underlying truth. Mathematical statements and schematic diagrams can highlight a problem's abstract principles, but when students get locked into regarding them as real, they obscure rather than clarify the underlying math.[7]

Although mathematics often deals in familiar quantities from daily life, it also encompasses unknown and variable values, where arbitrary symbols may stand for

"Just a darn minute — yesterday
you said that X equals **two**!"

some quantity I haven't figured out yet or a value that depends on changing conditions. Mathematics also deals in quantities for which one has no concrete frame of reference and can only barely imagine—i or ∞, for example. Thus, to do mathematics, one must adopt a hypothetical attitude, or an ability to distance oneself from the literal statement (*A is B*) and enter the abstract realm of imagination (*What if A were B?*). Appreciating symbols' arbitrary and abstract quality is part of what educator Abraham Arcavi has called "symbol sense."[8]

Rising above concrete experience to describe or draw conclusions about it represents one of the most difficult psychological challenges math students face. This section presents two examples of how students struggle to establish and maintain mental distance from the literal and concrete aspects of problems that often blind them to the underlying abstract mathematical concepts.

Word Problems

From first grade through college, one assignment most likely to trigger groans and eye-rolling is the word problem. Some of the linguistic difficulties that story problems can present have already been discussed; this section considers the larger conceptual speed bumps.

One role of word problems in the math curriculum is to serve as a proxy for, and a step toward, solving real-life problems. In textbooks and classrooms, however, word problems are also used the other way around—to illustrate general mathematical principles. In this function, they are merely cover stories.[9] (This applies to iconic illustrations, as well, such as pictures of real objects or situations.) For example, whether Matt and Jaime are collecting marbles, or Jason and Julie are collecting ribbons, the point of those particular stories is usually to demonstrate the principle and application of addition. Indeed, most specific information in word problems (other than the quantities and their relationships) is window dressing. Consider, for instance, the following problem:

> Stephie has four apples, two pears, and three marbles. How many
> pieces of fruit does Stephie have?

In this problem, the three marbles are irrelevant, of course: They are not fruit. Stephie's name is also mathematically irrelevant, as is the particular nature of her belongings, except that two of her collections are related by category. Strictly speaking, the problem is about summing four elements of one set and two elements of another, both subsets of a more abstractly defined third set. Students must figure this out before they even get to the arithmetic.

To make matters even more complicated, students must also sort through the cover story to determine whether the information in it is adequate for the mathematical solution. An example of a problem with inadequate information is the following:

> Thomas drove to visit his ailing Aunt Tillie 50 miles away. He left his
> house at 9:00 in the morning. What time did he arrive at his aunt's
> house?

In this problem, the bare mathematical bones consist of distance and time, which are related by rate, the critical missing element. Neither Aunt Tillie, nor her state of health, nor her solicitous nephew is mathematically relevant.

Cover stories can be both a blessing and a curse. On the plus side of the ledger, cover stories are much more entertaining and engaging than their no-frills mathematical counterparts. Students can often draw on some of the stories' surface characteristics—the objects, context, phrasing, story line, and the like—to recognize their underlying principles. For example, *altogether* frequently cues addition or multiplication, strings of similar objects (apples, pears, and oranges, say) usually signal addition or subtraction, and dissimilar but related objects (apples and baskets) commonly prompt division. This predictability is sometimes useful because it can lead to short cuts and more efficient solutions on similar problems. A recent survey of the word problems in a popular elementary textbook series found that, almost without exception, the objects and their relationships were aligned as expected with the desired operations.[10] More important, these regularities are pedagogically valuable because they mirror real-life situations and concepts children already understand. One can certainly add apples and baskets, but doing so will not necessarily further the goal of helping young students solve real problems. Students' real-world knowledge, such as knowing apples and pears are both fruit, should play a role in solving textbook problems, just as it should serve as a check against nonsensical results, such as a recipe calling for three quarters of an egg or a school trip requiring half a bus.[11]

The disadvantage of cover stories is that, when approached mindlessly, they can preclude insight into the problems' mathematical structure. For example, although one frequently does have occasion to add apples and pears, one would not use addition to find their ratio. Students' inability to rise above the irrelevant details can blind them to the question's mathematical structure and impede problem solving. Studies have shown that bilingual students and those with intellectual disabilities, emotional troubles, or specific learning disorders have particular difficulty sorting out the mathematically necessary information.[12]

Cover stories can create special difficulties when students use one word problem as an analogy to solve another. Students encounter this situation frequently in textbooks that offer a sample problem in the text and a series of related practice problems at the end of the chapter. For example, investigators found that students had difficulty solving permutation problems that assigned set A to a dissimilar set B (e.g., prizes to students) when they had been instructed on an example of simple pairing, in which the "assigned" set (e.g., students in one class) was concretely similar to the set being assigned to (e.g., their buddies in another class). They also had difficulty understanding that a problem involving the familiar situation of assigning prizes to students was mathematically similar to one that assigned students to prizes, a less familiar scenario.[13] In these examples, the students' reliance on their everyday knowledge of the cover story's mathematically irrelevant features (here, the sets' concrete similarity or the scenario's familiarity) impeded their ability to see below the surface to the problems' analogous mathematical structures.

One oft-cited and notoriously vexing problem is the following:

> Write an equation using the variables S and P to represent the statement: "There are 6 times as many students as professors."

In a series of studies, the question was answered incorrectly by more than half of the students in a college algebra class, by a similar proportion of freshman engineering students, and by more than a third of English-speaking engineering undergraduates and calculus-level students. The most common error was $6S = P$. (The correct answer is $6P = S$.) Although most of the students who produced the reversal understood that there were more students than professors, they could not extricate themselves from the literal statement and had apparently used S and P concretely to represent not the quantity of students and professors, but the students and professors themselves: six students, one professor (and an apparently meaningless equals sign). Even after researchers provided a specific lesson—demonstrating that if $S > P$, then P, not S, must be multiplied by 6 to make a true statement—the students continued to produce reversals, leading the researchers to conclude that their misconception (one might say their literalness or concreteness) was remarkably resilient.[14]

Students performed better on a more familiar-sounding version of the same problem:

> At a certain university there are 3,450 students. If there are six times as many students as professors, how many professors are there?

The advantage was strictly limited, however. The same individuals performed poorly on a subsequent, mathematically analogous problem in which they were asked to model the number of nails and screws produced by a factory, where there was no expected relationship between the quantities of nails and screws.[15] Thus the familiarity helped the students in only a very narrow sense; it did not help them generalize their knowledge to superficially different but mathematically identical problems.

Is reasoning by analogy necessarily difficult? Researchers have observed that in real life (e.g., scientific deliberations, political and journalistic rhetoric) people successfully argue by analogy all the time. Moreover, such analogies are often quite far flung. One political example likened Quebec's possible secession from Canada to surgery or to jumping into a lifeboat. What makes these circumstances so different from the math classroom? One important difference is that in these situations, people are spontaneously generating their own analogies. They are striving for a more effective way to argue a point and in so doing are attending to meaning, not superficial features. In the classroom and in textbooks, by contrast, the arguments and analogies are handed to students. In a further experiment, the same investigators from the study described above asked people to relate novel stories to model ones. In one condition, the participants just read the model story; in the second condition, they were asked to read it and then to make up their own analogy to it. Those people who were challenged to create their own analogies developed a much firmer grasp of the model and could much more successfully draw on it later to help them understand the novel stories.[16] How this method may apply to the mathematics classroom will be discussed in the Classroom Implications section.

Thinking in More than One Dimension

> All right, everybody line up alphabetically according to your height.
>
> Casey Stengel[17]

Few mathematical tasks strain students' capacity for abstract thinking as much as trying to wrap one's mind around more than one dimension at a time. Many such problems lie in wait to ambush the unsuspecting math student. Consider, for example, the question of what happens to the area of a square if one doubles its perimeter. Studies have revealed a deep-seated tendency among young adolescents to overgeneralize the principles of linear proportionality, such that they think that doubling the perimeter would also double the area. This stubborn misconception and its close cousin, the notion that shapes with the same perimeter must have the same area, has been called the *illusion of linearity*.[18]

Not all dimensions are spatial; in real-life problems, objects or events may have several different ordered characteristics to be considered simultaneously (e.g., price and time, age and weight). One method mathematicians use to analyze the relationship between such linked dimensions is Cartesian coordinates, the familiar perpendicular number lines intersecting at their origins, with each dimension's ordered values represented on an axis. When a value is plotted on a single number line, its position serves as an analog of its value. The points plotted in the Cartesian system, by contrast, represent two or three dimensions' values simultaneously, and most resulting graphs are *not* analogs. Such graphs are abstract schematics, not concrete pictures of the relationships they describe. Figure 9.1 abstractly depicts the motion of an object as follows: The object remains in one position for awhile, then moves at a constant rate, and then is still again for awhile. Seeing the graph as a concrete picture of motion, however, one might mistakenly interpret it as the path of a ball rolling along a table and then down a slide and along the floor.

When students first learn to graph functions on Cartesian coordinates, they frequently overgeneralize the linear analog concept to the resulting multidimensional graph. Some researchers have attributed this overgeneralization to earlier point-plotting exercises in which the connected dots actually do form pictures. Without proper instruction, this graph-as-picture misconception often goes undetected and uncorrected, persisting into adulthood. Artists and others who prefer a visual approach to problem solving, have weak spatial ability, and have a strong iconic imagery style are especially vulnerable, although even science students are not immune. By contrast, students with strong spatial ability and a preference for spatial imagery tend to interpret graphs abstractly. Interestingly, researchers found that when students with either visual style (spatial or iconic) attempted such problems using formal calculations, they trusted their visual representation, right or wrong, more than their notational one.[19]

Mathematics, at its core, is the most abstract intellectual endeavor, yet students' thinking often gets stuck on the concrete and literal features of problem cover stories and other representational conventions, obscuring the problems' abstract mathematical structure and significance. Thus one of teachers' greatest challenges is to demonstrate the transparency of mathematical words and symbols and help students see through them to the concepts they represent.

Development

It is commonly assumed that the demand for abstract thinking in math education begins with the jump from arithmetic to algebra. Abstract thought plays an important role very early in a child's mathematical education, however, beginning with counting. Although all the counting rules are generally applicable, and hence abstract, we focus here on the principle that any discrete, real or imagined object or event can be counted—a concept that derives from the fact that numerosity is independent of objects' or events' other properties.

Infants' numerical abstracting abilities are hotly debated. Some developmental psychologists have found evidence that infants appreciate the equivalence between three objects and three sounds, for example.[20] Others, however, have argued that

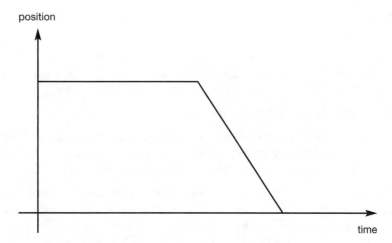

position

time

Figure 9.1. Students were asked to describe the situation represented by this graph. (From Kozhevnikov, M., Kosslyn, S., & Shephard, J. [2005]. Spatial versus object visualizers: A new characterization of visual cognitive style. *Memory & Cognition, 33,* 721; reprinted by permission.)

children gain this insight only gradually; here we examine the evidence for this more conservative position. In a series of experiments, 2½-year-old toddlers were unable to match two same-size sets, even when the objects in the two sets were nearly identical (e.g., dots and disks). They could not yet see that the three dots were equivalent to the three disks—to comprehend the abstract "three-ness" connecting the two groups. Over the course of the preschool years, however, children were increasingly able to match sets, or abstract numerosity, across a broader range of objects (e.g., dots and disks, then dots and shells), such that by age 4½ or 5 years, they could reliably match, compare, and add sets of dots with sets of sounds. Thus, over the course of about 2 years, children move from a concrete understanding of numerosity, whereby number is attached to particular objects, to an abstract one, in which number is independent of the objects it describes. Although these experiments were nonverbal, only those children with at least some verbal counting proficiency were able to move beyond the concrete point of view—once again highlighting the importance of language to mathematics and conceptual development.[21] By adulthood, of course, people can easily and accurately approximate and compare numerosities of all kinds of things and events, confirming that most adults have a firm sense of number's abstract nature.[22]

Some educators have noted that students seem to take a step backward in their abstracting ability when introduced to new concepts or notation. For example, children who as preschoolers and kindergartners had little trouble analyzing problems and improvising strategies for solving them, suddenly as first and second graders, using their newly learned formal rules and notation, find themselves rigidly wed to a single strategy for each operation.[23] In learning novel mathematical notation, as perhaps in learning to read a second language, students may adhere to the literal text until they have a feel for the new register. Indeed, in the same way, many bilingual students new to English struggle with abstract and idiomatic references in math text.[24]

One can also see the protracted developmental course of abstract thinking in students' performance on word problems. Researchers gave two sets of problems to groups of middle school and high school students. The first series was worded concretely:

> A young farmer has eight more hens than dogs. Since hens have two legs each, but dogs have four legs each, all together the animals have 118 legs. How many dogs does the young farmer own?

The second series was worded abstractly:

> The value of a given number is six more than the value of a second number. The sum of two times the first number and four times the second number is 126. What is the value of the second number?

All the students performed better on the concrete problems than on the abstract ones, although the performance gap closed somewhat among the older students.[25] It is not clear from this example, however, what made the farmer problem easier for students. Was it the familiar context? If so, at what point in development can students get past the cover story? From a mathematical point of view, one can solve a hens-and-dogs problem just as easily and validly (albeit nonsensically) with four-legged hens and two-legged dogs. Testing students on both concrete variations might help pinpoint that developmental milestone.

Logical Inference and Hypothesis Testing

This section ventures into the heart of mathematical reasoning to discuss logical inference and hypothesis testing—the truth-seeking thought experiments of mathematical science. People rightly associate logic with the algebraic solutions and geometric proofs of high school math. Like abstract thinking, however, logical reasoning pervades mathematics, even in the most basic arithmetic, and the skill has been statistically linked to math performance from the earliest grades.[26] The two principle methods of logical inquiry used in basic mathematics are deductive and inductive reasoning.

Deductive Reasoning

In deductive (or syllogistic) reasoning, one starts with premises assumed or declared to be true and draws conclusions from them. Consider this (nonmath) example:

> If all feathered creatures are birds, and sparrows have feathers, then sparrows are birds.

This is the simplest example of a syllogism—the most common mode of reasoning in formal mathematics and "the mortar that holds mathematical ideas together."[27] Deductive reasoning is a top-down method for arriving at certain truth, in which one's conclusions will necessarily be true if the premises are. Conversely, if one observes evidence that runs counter to the conclusions, the premises must be revised. In this sense, deductive inferences depend solely on an argument's internal consistency. In the mathematics classroom, algebraic solutions and geometric proofs are based on deductive argument, where to show one's work means to make one's case. More broadly, however, it undergirds all conventional math in so far as one starts with basic truths or definitions and derives other information from them. In even the simplest arithmetic problem, one begins with a mathematical premise (e.g., *If four plus two equals six . . .*) and a condition (*Keisha has four apples and two pears . . .*),

Figure 9.2. Deductive logic problem. A) Standard phrasing. B) Rephrased. The answer to both is Not Enough Clues. (From O'Brien, T.C., Shapiro, B.J., & Reali, N.C. [1971]. Logical thinking—Language and context. *Educational Studies in Mathematics, 4,* 213–214; reprinted by permission.)

from which one draws a logical inference (*then Keisha has six pieces of fruit*). Because it is possible that the premise is not true, however, such reasoning requires one to maintain a skeptical attitude—to regard all conclusions as being no stronger than the assumptions upon which they depend, which is a healthy approach to most intellectual endeavors.

Scholars have debated the verbal and nonverbal contributions to deductive logic. Some research has highlighted the importance of language. For example, the brain activates during deductive reasoning in the very language circuits involved in both phonological working memory and syntax.[28] Moreover, the conventional language of deductive statements can frequently impede, rather than enhance, understanding. For example, in a study of children in grades 4–10, researchers tested logical inference using both *if . . . then* language and a rephrasing in dialogue format, as illustrated in Figure 9.2. The investigators found significantly better performance at all grade levels with the rephrased propositions.[29] Readability assessments of both general and mathematical text have shown that statements held together by logical connectives are among the most difficult to read, understand, and construct. Some such phrases that are frequently used in math problems introduce a hypothetical premise (*if, suppose, let's say, given that, such that, so that, if and only if*). Others link premise to conclusion in forward (*then, consequently, thus*) or reverse (*because*), whereas still others highlight contradiction (*either . . . or, but, however, nevertheless*). One of the most difficult constructions is the negated premise (e.g., *Not all birds fly . . ., Not all integers are positive . . .*). Because students tend to interpret these statements literally (whereby *not all = some*), they often arrive at the wrong logical conclusion (*some birds fly*) rather than the correct conclusion that *some birds do not fly*. Research has tied verbal logical reasoning to reading ability; thus, although many students find propositional language troublesome, children with reading disabilities may find it especially

Five men were in a hospital. Each one is suffering from a different disease.

1. The man with asthma is in room 101.
2. Mr. Alex has cancer.
3. Mr. Osborn is in Room 105.
4. Mr. Wilson has TB.
5. The man with mono is in Room 104.
6. Mr. Thomas is in Room 101.
7. Mr. Wilson is in Room 102.
8. One of the men has epilepsy.
9. One of the patients is in Room 103.

What disease does Mr. Young have?

Name	Disease	Room
1.	Asthma	101
2. Mr. Alex	Cancer	
3. Mr. Osborn		105
4. Mr. Wilson	TB	
5.	Mono	104
6. Mr. Thomas		101
7. Mr. Wilson		102
8.	Epilepsy	
9.		103

Figure 9.3. Deductive logic problem, solved using a matrix. (From Schwartz, S.H. [1971]. Modes of representation and problem solving: Well evolved is half solved. *Journal of Experimental Psychology, 91,* 347 [published by the American Psychological Association]; Schwartz, S.H., & Fattaleh, D.L. [1972]. Representation in deductive problem solving: The matrix. *Journal of Experimental Psychology, 95,* 344 [published by the American Psychological Association].)

difficult.[30] Propositional logic has also proven vexing for students learning English as a second language, particularly students whose first language is Spanish, in which two negatives denote a negative.[31]

Other research has highlighted nonverbal contributions to deductive logic. One's capacity for grasping cause and effect in everyday life does not necessarily depend on language and is governed by a different brain region from that governing syntax.[32] Both brain and behavior studies have provided evidence that one can create visual schematic images to track and analyze logical chains of conditions, relationships, and inferences.[33] Indeed, schematic images are an effective and efficient method for representing many deductive reasoning problems, particularly because they significantly reduce the verbal working-memory load, as in the matrix illustrated in Figure 9.3.[34]

Not surprisingly, students' performance on verbally or schematically represented problems varies with their abilities in those respective domains.[35] One potentially fruitful line for future research would be to examine the effect of problem-solving style on deductive reasoning performance. Iconic visualizers, for example, may have particular difficulty representing the systematic chain of logical argument.

Inductive Reasoning

Inductive reasoning, in contrast to deductive logic, is a bottom-up method of arriving not at truth exactly, but at the *probability* that something is so. For example, suppose that a person observes that every bird he or she sees flies. One could most reasonably come to the conclusion, not that *all* birds fly, but that *most birds probably fly*; a visit to the zoo's ostrich pen would reinforce this cautious interpretation of the data. Unlike deductive argument, inductive reasoning depends entirely on real-world observations and the probabilistic inferences that a person can draw from them; as such, it forms the basis of the scientific method and is modeled mathematically by statistics.[36]

Earlier in this chapter, we discussed solving novel problems by analogy to familiar ones. Thinking by analogy depends on inductive reasoning: When the accumulated evidence suggests that one problem is mathematically the same as another, it is likely that the representation and solution for one will work for the other, as well. Suddenly recognizing a familiar problem type often prompts the exclamation that teachers love to hear, "Oh, yeah! I know how to do that!" Reasoning by analogy, like inductive reasoning in general, depends on a person's store of applicable experience. It also relies on the flexibility, fluidity, and sometimes creativity of a person's associations to those models, which in turn depend on one's grasp of them. Writing of scientists, psychologist Usha Goswami noted,

> Scientific breakthroughs often depend on the right analogy . . . but the scientists who make the breakthroughs seldom have extra information that is unavailable to their colleagues. Instead, the analogy occurs to them and not to their fellow scientists because of the way that their conceptual understanding of their field is structured, and the richness of their representations.[37]

This characterization may apply to math students as well, some of whom struggle not because they lack appropriate prior experience to serve as models, but because they do not fully understand those models and thus do not spontaneously think of them.

Most learning depends on inductive reasoning. In math lessons, repeated exercises that demonstrate a concept give students the opportunity to amass the data they need to develop a mental category, not only of the problem type but also of the sort of information one needs to solve each kind. This way, students stock their inventory of useful information and experience from which they can draw in the future. Although formal proofs and solutions depend upon deductive reasoning, much of the mental effort leading up to producing the formal argument is messily inductive and to be found in the scrap paper scribbles, not in what is submitted when asked to "show your work." When confronted with a puzzle, the mathematician must gather all the data—the given values, the conditions and premises, the question being posed, and whatever mathematical information this problem may bring to mind—spread them out, discard the pieces that do not seem relevant, keep those that are, and by educated experimentation try to fit them together. As noted mathematician George Pólya observed,

> [B]efore obtaining certainty we must often be satisfied with a more or less plausible guess. Without considerations [that] are only plausible and provisional, we could never find the solution which is certain and final. . . . Mathematics presented with rigor is a systematic deductive science but mathematics in the making is an experimental inductive science.[38]

Testing Hypotheses

Both deductive and inductive logical models allow one to make certain testable predictions (hypotheses) about things and events, and to assess the validity of one's conclusions. This kind of thinking demands a great deal of working memory and attention, plus easy access to well-learned mathematical information established in long-term memory. It also requires sequencing and anticipation to link the premise or observation to the conclusion, as well as a hypothetical attitude, or the willingness to consider premises and inferences that could—but may not—be true. Looking at problems from various angles and dropping ideas if they do not pan out require mental flexibility. Verifying one's ideas is exceptionally important in mathematics and is an extension of the self-monitoring skills discussed previously.

The difference between the logical methods has important implications for the way reasoning tasks are approached in the mathematics classroom. It has been suggested that the so-called math wars have evolved from the fundamental differences between an analytical, deductive approach to truth on the one hand and an empirical, inductive approach on the other, frequently dubbed "back-to-basics" and "contructivism," respectively.[39] Clearly, both methods are essential to math education.

Development

A child's logic is chiefly inductive, which simply means that children learn from experience. Previous chapters discussed how infants use their experience to predict the motion of objects and how preschoolers learn about counting from everyday events. Children develop a set of expectations and theories about why things happen based on past experience, and then refine those ideas when they get surprising new information.

Piaget's theories of cognitive development held that a child's capacity for logical thought arises only gradually from action and experience.[40] More recent studies of inductive reasoning have revealed substantial skill among even very young children. For example, verbal analogy problems (*sparrow:flying::ostrich:?*), which were the bane of high school SAT-takers until recently, require a high level of verbal inductive reasoning. One study has shown, however, that children as young as 3 or 4 years can solve A:B::C:D analogies in *pictorial* form (Figure 9.4), provided they understand the relevant relationship between A and B. These findings prompted the researchers to conclude that inductive reasoning comes naturally to young children, and that any development in that skill simply reflects the fact that they come to know more over time.[41]

Piaget's developmental theories also posited that around age 11 or so, children begin to be capable of the abstract thought experiments of deductive logic.[42] In a series of elegant experiments, however, Thomas O'Brien and his colleagues revealed that the developmental trajectory of formal, deductive logic is more complex than was originally believed. The researchers investigated the development of deductive reasoning between 4th and 10th grades. As illustrated in Figure 9.5, they asked the children to think of a pile of colored chips and set forth the premise: *If I show black, you show white.* They then stated a condition and inquired about the implication, for which the children were to choose one of the following answers: *yes, no,* or *not enough clues.* One question, for example, was *I show black. Do you show white?* The answer, of course, is *yes;* the question is a direct restatement of the premise. Another question was the counterexample: *You do not show white. Did I show black?* Previous research had shown that children as young as 7 years could easily (and correctly) answer *yes* to the first question and *no* to the second, suggesting that children younger than 11 were capable of some deductive thinking.

There are two other logical questions based on the original premise, however: the inverse (*I do not show black. Do you show white?*) and the converse (*You show white.*

Figure 9.4. Example from a picture analogy test given to 4-, 5-, and 9-year-old children. Answer choices include D) correct analogy, E) thematic association to C, F) category association to C, and G) mere appearance match to C. Most children at all ages choose D. (Reprinted from *Cognition*, Vol. 36. U. Goswami and A.L. Brown, Higher-order structure and relational reasoning: Contrasting analogical and thematic relations, page 211. Copyright 1990, with permission from Elsevier.)

Did I show black?). These questions are significantly trickier. Upon reflection, one realizes that the correct answer to both questions is *not enough clues*. Yet nearly half of all the children in each grade in the study consistently answered *no* to the first question; three quarters of them consistently answered *yes* to the second. That is, they seemed to understand the premise to mean, *You show white if and only if I show black*, thereby erroneously precluding the possibility of your showing white when I do *not* show black.

The results were similar when researchers repeated the study with students in grades 4, 6, 8, 10, and 12, as well as with college undergraduates who had just completed a logic course. In the replication studies, the researchers used real-life sorts of premises (e.g., *If Joe will play, the Tigers will win; If the car is red, it is Joe's*). Fewer than 10% of the students in the studies' lowest grades consistently responded correctly to inverse and converse items—a ratio that did not significantly improve with development into early adulthood, despite the students' average to above average intelligence (and, in the college group, despite specific instruction). Rephrasing the question to eliminate *if . . . then* language made the task somewhat easier; in that case, the older children performed better than the younger ones. Even with rephrasing, however, only half of the 10th graders consistently responded correctly to inverse and converse problems, suggesting that *if . . . then* language was not the only source of their reasoning difficulty.[43]

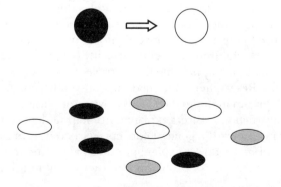

Figure 9.5. "If I show black, you show white." (*Source:* O'Brien, Shapiro, & Reali, 1971.)

Thinking logically and abstractly is exceptionally difficult for many children, and students' misconceptions have proven remarkably resistant to correction. As noted in Chapter 1, these higher-order mental skills constitute the core of basic intelligence. This takes the discussion back to the question asked in Chapter 3 regarding spatial skill: Can such fundamental skills as logical and abstract thinking be taught? To find out, we turn our attention to the classroom.

Classroom Implications

The more that students can think abstractly and reason logically, the more likely they will be able to generalize what they learn in school to a range of novel problems inside and outside the classroom—the point, many would agree, of education. Unschooled individuals often master isolated math skills necessary for work or other activities, but are ill equipped to apply them to problems outside that narrow range. At its finest, math education fosters reasoning and analytical skills applicable to problems well beyond the scope of mathematics proper. Educators continue to debate, however, whether and to what extent it is possible to ensure that students take generally serviceable problem-solving skills with them when they graduate. This section reviews the work of some of the more optimistic scholars in the field, who have produced just a few small studies and some general pedagogical suggestions.[44] The ideas are organized into six basic pedagogical principles.

1. *Provide explicit explanations,* or old-fashioned explanatory instruction. In one study, researchers showed children, ages 5 to 10, a page with drawings of six dogs and three cats and asked, "Are there more dogs or more animals in the picture?" All the children could count correctly, but most of them answered as if they were comparing the number of dogs to the number of cats. Pointing out to them that there were six dogs and nine animals, for example, helped some of them realize their error, but the teacher's explanation that dogs are a type of animal—the abstraction linking dogs and cats—proved a much more effective intervention that led to prolonged and generalized improvement in the children's ability to solve similar problems, even for children as young as 5. In addition, most of the students thought the second explanation was "smarter" than the first, even if they did not quite understand it right away. The authors concluded that "when young children understand that logical solutions are possible and know how to apply them, they readily adopt them."[45]

   In another study, in which researchers spent several months giving problem-solving tips to third graders, investigators found that the more information and strategies the children had been explicitly taught and had at their disposal, the more versatile and applicable their problem solving was, regardless of their ability level.[46] Researchers obtained similar results comparing college students doing their own in-depth data analysis with and without the benefit of additional expert instruction.[47] Direct explanations help students integrate, put into words, and set in context the general concepts garnered from firsthand experience. In the case of math, direct explanations also alert students to the mathematically relevant dimensions of problems that they might not think about otherwise.

2. *Analyze and anticipate faulty assumptions.* One common source of conceptual error is a faulty assumption. Research is needed to determine whether students can effectively be taught to question the concrete preconceptions, often imported

from daily experience, that interfere with grasping mathematical concepts. It would also be useful to know whether identifying and challenging other, more general, distracting assumptions at work in the classroom (and life) improve problem solving. These assumptions include the expectation that the given information is relevant and adequate, that easy problems appear first, that correct answers come out even, and that all problems have solutions.[48] Investigators should also test the effectiveness of explicitly warning students about the most common pitfalls, including using "buggy" algorithms (see Chapter 5), interpreting graphs as pictures, or assuming *if* means *if and only if.* Would such interventions prompt students to dig deeper?

3. *Preview and review problems.* Students often feel that arriving at the correct answer is more important than understanding the problem. One potentially effective teaching strategy is to examine the problem thoroughly, perhaps without even solving it at all. What is the problem asking? How could one restate the problem? What type of problem is it? What does one need to know to solve it, and does one have that information?[49] What is the best procedure to use? What are other ways it could be solved? Is there an appropriate standard formula, and why does it apply? Why would one rule out using other formulas? What kind of answer does one expect to get? What is the approximate value of the expected answer?[50] Are there familiar patterns that can help? If one does solve the problem, how does the answer compare with the predicted answer; if there is a significant discrepancy, what is the reason for it? How would solving this problem help one solve others? An important aspect of previewing a problem is determining how best to represent it (e.g., by equation, diagram, matrix, graph)—a process that should focus a student's attention on its mathematical structure. Given representations often apply to a broad range of problems, even though the problems may require different solution strategies. Experience with a representation makes it more likely that a student will use it spontaneously and appropriately on different types of problems.[51]

4. *Provide examples.* Students new to a topic learn better from studying worked examples than from solving their own problems without this support. This method, along with individual instruction, is especially useful for struggling students and ensures that they do not just practice their own errors and misconceptions. As students become more proficient, they outgrow their need for this support.[52] One way to help students discern which problem features are solution-relevant is to have them note the mathematical similarities between multiple analogous problems with disparate cover stories—a method psychologist Miriam Bassok calls "abstraction by intersection."[53]

Although many textbooks offer a broad range of analogous problems at the end of each chapter, students sometimes miss the relevance as they focus on obtaining the correct answer and completing the assignment. The problems' structural similarities are worth pointing out explicitly in class. Seeing one concept in action in different settings and different principles in play in superficially similar settings allows the student to develop a deeper understanding of the underlying mathematics. For example, students who have completed a trigonometry class have a better differentiated understanding of right triangles than does a beginning geometry student because they have solved a broader range of problems using right triangles. Solving a range of problems also builds a web of associations that will help the student recall mathematically, rather than superficially, analogous models for solving new problems.[54]

5. *Real-world sleuthing.* Researchers instructed fifth graders of average ability using a series of videos depicting a real-life problem with the quantitative data embedded naturalistically in the story; for comparison, they gave another group of students instruction in the same problems in standard word-problem format. They found that the students taught with video simulations later performed significantly better on more complex simulation problems when compared with students taught in the traditional way. These results led the researchers to conclude that having an opportunity to identify the problem and relevant data and devise a solution for themselves gave students richer insight into problems and heightened their problem-solving ability.[55]

6. *Encourage students to generate their own problems, data, and formulas.* The educational literature is replete with calls for students to take an active role in their own learning; math class certainly offers many such opportunities. One study examined the effect of writing original "math stories" and deriving problems from them on third- and fourth-grade students' problem-solving skills. With explicit instruction, the students, who had a broad range of ability levels, learned to create stories based on different models and to combine and vary them. Well past the end of the program, their problem-solving skills continued to surpass those of children who had been given guidelines and ample practice but no chance to create their own analogs.[56]

   Students also gain insight by controlling the data. Several studies have demonstrated success in teaching graphing (and in correcting the graph-as-picture misconception) by using computers and sensory probes to control and track changes in such things as temperature, light, and force in real time on computer-generated graphs in the physics laboratory. Other studies have shown that student-generated data on graphing calculators furthered students' grasp of the nature of variables and basic functions. Unfortunately, none of these studies were adequately controlled, so it is uncertain whether these techniques are truly superior to any others.[57]

   Another study showed that students developed a rich appreciation of variance by exploring several data sets and developing their own formulas to describe variability. They also better understood the conventional formulas, which were taught subsequently.[58] The pitfall of introducing concepts with the students' own ideas, of course, is that the students may "learn" their misconceptions. Therefore, immediate feedback and follow-up instruction is essential to success with this pedagogical strategy.

Do these methods work? Not always. For example, one fine effort to correct eighth graders' illusion of linearity used several of these strategies, such as explicitly addressing students' preconception and providing appropriate mathematical models. Nevertheless, the interventions produced disappointing results: Students made some improvement, but also made new types of conceptual or technical errors and overgeneralized nonlinear responses to linear questions.[59] The number of studies on these methods is very small and many questions remain unanswered. For instance, do concrete or abstract examples work best for enhancing abstract thinking and application of learning? One study found that concrete story problems worked best for introducing basic algebraic concepts, but that abstract equations worked better for more complex ideas.[60] The scholarly debate over the relative value of concrete versus abstract examples, however, remains lively.[61]

It is also not yet clear whether logical reasoning can be improved and, if so, by what method.[62] Chapter 2 discussed the value of certain types of board games in teaching number sense. Does playing logic games (e.g., Mastermind, Clue, Sudoku) enhance deductive reasoning, and will students bring those skills with them to math class? How should methods be modified, if at all, for younger or older children? One could expect that the natural maturation of attention, working memory, and planning would strongly influence students' capacity for abstract thinking and logical argument. Specific research is needed to sort this out.

Most of the studies cited here involved children of average ability. It is hard to say to what extent these techniques are helpful to children with specific cognitive or perceptual impairments. For example, can a child with spatial difficulties benefit from diagramming a problem or using a graphing calculator to understand algebraic functions? To what extent can children with limited imagination benefit from analogies, understand suppositions, or reflect on their own assumptions? Do individual children's developmental trajectories vary, or are some children naturally more limited than others in the ability to abstract? Given the close association of these functions to intelligence, one may reasonably ask whether all students will necessarily benefit equally from instruction geared to abstract concepts and complex logical argument. More broadly, how fixed is intelligence and how far can it be stretched?

Finally, pedagogical research has not shown how long children retain their insights; many studies do not look beyond a day or a week. How do they solve problems 10 or 20 years after graduation? In the lifespan study of high school mathematics retention cited in Chapter 5, the researchers suggested that math's inherent conceptual coherence and organization make it less vulnerable to long-term decay than other domains, such as foreign-language vocabulary.[63] This bulwark, however, depends on students' conceptual grasp of the material. A sensible math curriculum will need to consider long-range goals.[64] More research on teaching children of all ages and ability levels to reason in various mathematical contexts is clearly warranted.

Conclusion

Reasoning lies at the heart of mathematical thought. To solve any mathematical problem, one must think abstractly and logically. The developmental trajectory of these skills extends from very early childhood well into adolescence, affecting all math topics. Some of these skills are challenged most severely at certain critical learning points, such as the introduction of new concepts. Compromised reasoning is most striking among students with limited general intelligence (by definition, as discussed in Chapter 1), as well as among children with specific learning or emotional difficulties. At the present time, scholars have only a rudimentary idea about whether and how these fundamental skills can be taught and whether improvements in them would bear fruit in the mathematics classroom.

Notes

[1]Pólya, 1945, p. 181.
[2]Mahone et al., 2002.
[3]Cattell, 1971; Mackintosh, 1998.

[4]Locke, 1690/1824.

[5]As cited in Halpern, 1992, p. 7.

[6]Pimm, 1987, p. 139.

[7]Presmeg, 1992.

[8]Arcavi, 1994.

[9]See Bassok, 2003, for the review upon which this section is based.

[10]Bassok, Chase, & Martin, 1998.

[11]Bassok, 2001.

[12]Parmar, Cawley, & Frazita, 1996.

[13]Bassok, Wu, & Olseth, 1995.

[14]Clement, Lochhead, & Monk, 1981.

[15]Bassok, 2001, pp. 416–421.

[16]Dunbar, 2001.

[17]Stengel, n.d. Quote of Casey Stengel provided by the Estate of Casey Stengel by CMG Worldwide, www.cmgworldwide.com.

[18]See, e.g., Van Dooren, DeBock, Hessels, Janssens, & Verschaffel, 2004; Dembo, Levin, & Siegler, 1997.

[19]Kozhevnikov, Hegarty, & Mayer, 2002; Kozhevnikov, Kosslyn, & Shephard, 2005; McDermott, Rosenquist, & van Zee, 1987.

[20]Kobayashi, Hiraki, & Hasegawa, 2005; Starkey, Spelke, & Gelman, 1990.

[21]Barth, LaMont, Lipton, & Spelke, 2005; Mix, 1999; see Mix, Huttenlocher, & Levine, 2002b, pp. 47–48, for a review and critique.

[22]Barth, Kanwisher, & Spelke, 2003.

[23]Carpenter, Hiebert, & Moser, 1983.

[24]Abedi & Lord, 2001; Spanos, Rhodes, Dale, & Crandall, 1988.

[25]Caldwell & Golin, 1987.

[26]Gregory & Osborne, 1975; Roberge & Flexer, 1983. See also Floyd, Evans, & McGrew, 2003, for the link between fluid reasoning and math in the Woodcock-Johnson (3rd ed.) psychoeducational test battery norming data.

[27]O'Brien, Shapiro, & Reali, 1971, p. 218.

[28]Goel & Dolan, 2004.

[29]O'Brien et al., 1971.

[30]Gregory & Osborne, 1975.

[31]Mestre, 1988; Spanos et al., 1988.

[32]Varley & Siegal, 2000.

[33]Dawe, 1983; Knauff, Mulack, Kassubek, Salih, & Greenlee, 2002; Kroger, Cohen, & Johnson-Laird, 2001, as cited in Knauff & Johnson-Laird, 2002; Varley & Siegal, 2000.

[34]Polich & Schwartz, 1974; Schwartz & Fattaleh, 1972.

[35]Sternberg & Weil, 1980.

[36]Goel & Dolan, 2004.

[37]Goswami, 2001, pp. 465–466.

[38]Pólya, 1945, pp. 146, 106.

[39]Campbell, 2005.

[40]Piaget, 1975.

[41]Goswami & Brown, 1990.

[42]Piaget, 1975.

[43]O'Brien, 1973; O'Brien et al., 1971; Shapiro & O'Brien, 1973.

[44]See Halpern, 1992; 1998, the principal sources for this section.

[45]Siegler & Svetina, 2006, p. 1010.

[46]Fuchs, Fuchs, Finelli, et al., 2006.

[47]Schwartz & Bransford, 1998.

[48]See, e.g., Boaler, 2000; Luchins, 1942; Reusser, 1988.

[49]Low, Over, Doolan, & Michell, 1994.

[50]Sowder, 1992.

[51]Novick, 1990.

[52]Carroll, 1994; Kalyuga, Chandler, Tuovinen, & Sweller, 2001.

[53]Bassok, 2003, p. 352.

[54]Schwartz & Bransford, 1998; Xin, Jitendra, & Deatline-Buchman, 2005.

[55]Van Haneghan et al., 1992.

[56]Rudnitsky, Ethridge, Freeman, & Gilbert, 1995.

[57]Ferrini-Mundy & Lauten, 1993; Graham & Thomas, 2000; Linn, Layman, & Nachmias, 1987; Mokros & Tinker, 1987.

[58]Schwartz & Moore, 1998.

[59]Van Dooren et al., 2004.

[60]Koedinger, Alibali, & Nathan, 2008.

[61]Kaminski, Sloutsky, & Heckler, 2008; Sowell, 1989; Uttal, Scudder, & DeLoache, 1997.

[62]Gregory & Osborne, 1975.

[63]Bahrick & Hall, 1991.

[64]See Barnett & Ceci, 2002.

Section IV

Professional Implications

F
or teachers, school psychologists, and other readers privileged with front-row seats to children's mathematical development, the foregoing chapters probably brought to mind particular students they have known. If that is the case, and if the research reported herein permitted new insights into those students' strengths and difficulties, then *Number Sense and Number Nonsense* has achieved its primary goal.

However, many readers may still have questions: What can be done to help students who struggle with math or to enhance the learning of those who are muddling through? Are there concrete, practical guidelines with applications for day-to-day work? What are the implications of the research to date?

Most of the research on math difficulties is in its infancy and for the most part cannot yet support concrete pedagogical recommendations. That does not mean, however, that it has nothing useful to say to practitioners. Section IV outlines some of those implications, many of which are in the form of important questions for the future.

Chapter 10 offers some general guidelines for a comprehensive psychoeducational evaluation designed to identify perceptual, cognitive, executive, and emotional contributions to a student's mathematical learning. Can one yet say for sure that a particular finding explains a particular difficulty? Unfortunately, the answer to that question is often no. But a careful and thoughtful assessment can shed light on the range of resources a student brings to the classroom, an invaluable asset to the front-line instructor. Chapter 11, the final chapter, summarizes the findings and outstanding questions relevant to teaching and then discusses a host of critical, unresolved issues currently facing mathematics education—including teacher preparation, curriculum, and pedagogy—that future research may one day be able to address.

Although the chapters in Section IV focus on matters of special interest to particular professions, they are not addressed to members of those professions exclusively. Teachers play a vital role in psychoeducational evaluation; psychologists and other professionals must understand how student, teacher, curriculum, and pedagogy interact in the classroom; and researchers must be aware of the practical questions if their work is to be applicable and relevant. All readers are therefore encouraged to consider both chapters.

Chapter 10

Evaluation

When students struggle in the classroom despite working hard, they are often referred for formal evaluation of learning disability. Many teachers, parents, and even psychologists have only a vague notion of what a learning disability is, but most understand that it involves a problem with reading and writing. Indeed, it is usually problems with reading and writing that prompt such a referral. As discussed in Chapter 1, even the legal definition of *math learning disability* is couched in terms of impairments in skills underlying written or spoken language. Only very rarely, if at all, will a parent or teacher ask for an evaluation solely or primarily for math troubles. If students struggle in math, they are just thought to be "weak in math." When such students are evaluated, the psychologist's report will usually simply describe the arithmetic facts and algorithms that the child does not know or will note poor reasoning on word problems. These findings are meant to alert the teacher about gaps in the child's mathematical knowledge, but rarely enlighten the teacher about the reason or reasons for failure.

It is not surprising, then, that few published guidelines are available for a comprehensive psychoeducational evaluation of mathematical difficulties.[1] (By contrast, there are volumes of professional books that delineate methods for evaluating learning disabilities as they relate to reading and language. That is not to say that there is a dearth of published diagnostic math tests.) Educators and researchers have devised a few excellent tests of mathematical skill based on scientific findings, reviewed later in this chapter. Those instruments go only so far, however. They can indicate that a student is struggling (or is at risk for later difficulty) and they show what types of mathematical tasks give students the most trouble—results that usually simply restate or confirm the teacher's observations. Although this information is certainly necessary, it is not sufficient.

Math is a highly complex mental enterprise associated with activity in multiple brain networks and the diverse perceptual, cognitive, executive, and higher intellectual skills those circuits support. Can an individual child's mathematical difficulties therefore be attributed to impairments in these mental abilities? That is a complicated question. Most of the studies designed to answer it are small and preliminary; thus, for many students, the "cause" of the student's math difficulties cannot be definitively determined. A comprehensive evaluation of the student's reasoning and perceptual, cognitive, and executive functioning, as well as of potential social and emotional contributions, however, will elucidate the resources he or she brings to the mathematics classroom. With that information, the teacher is in a better position to know what kinds of pedagogical methods are most likely to be fruitful.

The comprehensive evaluation outlined in this chapter aims to paint a psychological portrait of a student by placing his or her math skills in cognitive, behavioral, and curricular context. The results of this sort of evaluation will be descriptive rather than diagnostic. That is, they will not define a student as "having" or "not having" a math disability, and thus will not necessarily satisfy the need for sharp diagnostic distinctions as discussed in Chapter 1.[2] Rather, the assessment will describe the student's math skills and difficulties and speculate about the reasons for the latter. This approach is consistent with the scientific finding that there is no qualitative difference between the math difficulties of students with average and below-average abilities. The evaluation may even be useful for students with adequate math performance who have reason to believe they could do better. In all, it aims to provide insight for the classroom teacher.

This chapter outlines our suggestions for a comprehensive evaluation of a student's math skill and vulnerability. We begin by describing the wealth of information available even before testing begins and then outline the areas to be assessed by formal testing.

Informal Evaluation

The psychologist will be able to glean valuable information not only from talking with the student, but also from parents, teachers, work samples, grade cards, other school records, and prior psychoeducational evaluation reports. The examiner will first want to probe the concerns that prompted the evaluation.[3] The nature of these concerns will vary with the student's age and math level, but even with older students it will be important to inquire about basic skills: counting, comparing values, and recognizing numerals. Does the student grasp the concepts but struggle with computations? Or perform the computations but not see to what purpose? Do work samples reveal difficulties with writing mechanics, visual organization, carelessness, reading fluency, or comprehension? Does the student seem to lack a sense of numbers—are estimates and calculations markedly unreasonable? Does the student understand what the answer means? Can the student approach problems from different angles? Does the student prefer to work problems out verbally and symbolically or by drawing diagrams or even gesticulating? Are the student's math difficulties limited to the classroom, or do they affect his or her daily life (e.g., in reckoning time, money, or measurements)? What teaching methods has the teacher used, and has the student fared better with one particular method over another? Did the trouble begin this year or has it been going on for awhile? If the student has struggled for a long time, what is the reason for seeking an evaluation now? Did the onset of the student's difficulty coincide with pedagogical, curricular, or textbook changes?

The examiner will also want to inquire into difficulties in other areas. Some useful questions include the following. Does the student struggle with all homework or just some subjects or types of assignments? Conversely, are there subjects the student loves or excels in? How does the student feel about school? What elicits enthusiasm? Are reading, writing, and spelling at grade level? Are there signs of language disturbance? Can the student build towers with blocks, do jigsaw puzzles, read maps, construct models, or repair things?

The examiner will want to inquire about early development. Did prenatal or delivery problems place the student at risk? Were speech and motor milestones on time? Has the child been screened recently for hearing or vision difficulties, and are corrective lenses up to date? Which hand does the child typically use, and is it the same hand for all activities? Does the child get lost easily? Does the child know left from right? Historical perspective is important; all the information noted here as relevant for younger students should be obtained regarding older students' early childhood and schooling as well.

Some psychiatric disorders, such as depression and ADHD, are so common that it is wise to screen for them routinely. Longitudinal observations by parents and teachers, as well as the students' own complaints, will provide the most valuable diagnostic information in that regard.[4] Does the student seem distractible, repeatedly fail to hand in homework on time, have a disorganized backpack, or forget when tests are scheduled? Are the student's answers offered impulsively and do they seem careless? Does the student seem fidgety or play with pencils and other small objects? Are these behaviors new, episodic, or lifelong, and are they limited to math class or pervasive? What is the student's general mood? Does the student seem preoccupied, sad, or irritable? Have there been recent disruptions in the student's family or social circumstances?

Family history may be relevant. How far did the child's parents go in school? What are their occupations? What are their attitudes about the importance of, and potential for success in, mathematics? Do any close blood-relatives have learning difficulties, especially in math or reading, or have symptoms of ADHD?

The psychologist will want to inquire whether the child has ever been evaluated previously, and if so, what the findings were. Psychologists often find reviewing previous test scores informative: prior results sometimes take on new meaning in the light of later events or findings. Moreover, if the examiner has a choice of test instruments, electing the ones used previously affords a direct performance comparison over time.

Finally, the psychologist will want to hear the students' views of their math and general school experience. Some additional questions the examiner might ask the student include the following. What do you think the trouble is about, and what has learning math been like for you? What has come easily and what not so easily? What has the teacher said or done that has made things clear or more obscure? Do things go on in the classroom that make no sense at all? What useful techniques have you discovered for solving problems? Has math been harder than other subjects, or have other subjects also been difficult? In what way? Is it difficult to pay attention in math and other classes? Can you keep track of daily assignments? What do you like to do for fun? Do you read for pleasure? Did you ever enjoy playing with blocks and puzzles? Reading maps? Building or fixing things? Adolescents and young adults should be asked about driving: How do you navigate—by verbal instructions, landmarks, or maps? All students should be asked these broader questions: How do you feel about math, school, learning, and the importance of education for your life? What are your goals?

The examiner can also gain a great deal of vital information by simple observation. A student's scribbles and other behavior during formal testing can reveal clues to the student's problem-solving strategies. Other behavior includes finger counting, subvocalizing the number sequence, using tally marks, drawing pictures, gesturing, trial-and-error scribbling and erasing, and listing numbers. Does the student favor iconic or schematic drawings? Observation can also answer questions about executive functions. Does the student respond impulsively, fidget, or request the examiner to repeat questions? Are paper-and-pencil productions disorganized? Rapport-building conversation will yield information about speech, vocabulary, syntax, and comprehension. Finally, a student's testing behavior can reveal broader psychological information. Does the student try hard? Seem motivated? How does the student respond to frustration? What is the student's mood and attitude? Does the student seem self-confident regarding cognitive tasks in general and math and other subjects in particular?

The information gathered informally will allow the teacher and psychologist to develop hypotheses about the student's strengths and weaknesses, which, in turn, can help the psychologist to devise an efficient formal evaluation to test those hypotheses.

Formal Evaluation

Before we discuss the specifics of the formal evaluation, we offer several general points about testing. First, sound testing practice requires one to base interpretation on more than a single piece of evidence. Math tests should have multiple items for every skill tested; for every cognitive domain assessed, we recommend using at least two tests that approach the question from somewhat different angles.[5]

Second, the more focused a test, the stronger can be one's interpretation of the findings. If a particular test draws on several different skills—auditory processing, working memory, and inductive logic, for example—then it may be difficult to determine the reason for poor performance. No test is "pure," however, because all tests have perceptual and executive components, regardless of the cognitive demands. Individuals also often impose their own strategies, sometimes quite inventively, on even the most straightforward problems; the examiner will want to take note when a student uses an atypical approach to a problem.

Third, some currently available test instruments are very good and others are weak; for some skills, the ideal test has not yet been developed. In the following sections, we mention a few tests that may help answer the teacher's and psychologist's questions; we encourage psychologists to experiment with several to see which work best. One caution, however: Although the best tests for a given student may be subtests of unrelated test packages or batteries, the psychologist will need to be mindful of the difficulty in comparing results across tests that are standardized on different population samples and limit the number of batteries used.

Finally, math difficulties can sometimes be transient. In particular, for young children and for older students with no prior history of math difficulty, teachers and psychologists should monitor their performance over a period of time.

Achievement

The student's classroom portfolio, teacher's observations, and scores from nationally standardized, group-administered tests should suffice to determine whether the student has missed a significant portion of the curriculum or lags behind classmates, although the teacher or psychologist may want to verify those findings with a broad-based, individually administered test. To this end, there are several commercially available stand-alone math tests, such as KeyMath 3,[6] as well as general achievement batteries that include standardized measures of math performance. The most important analysis, however, will be a qualitative review of the student's responses, which may shed light on the reasons for math failure. For this purpose, an individually administered assessment will be necessary.

Basic Number Skills

Students with persistent and serious math underachievement often lack the most fundamental skills in either or both modes of quantitative understanding—the intuitive and the conventional—which must be surveyed in a comprehensive evaluation. Intuitive skills include estimating numerosities and comparing numeric values (e.g., "Which is bigger, nine or four?"). Students should be able to order values, locate them on age-appropriate number lines, and predict the effect of simple arithmetic transformations. Number sense also implies awareness of relative quantity in everyday life (e.g., "Would you pay $100 for an ice cream cone?"). Sometimes a student responds vaguely or seems bewildered when asked age-appropriate number-fact questions (e.g., "What is the boiling point of water?"), suggesting a general obliviousness to quantity in the environment. Conventional skills include counting in the age-appropriate range according to the governing principles and identifying ordinal rank. They also encompass reading, writing, and comprehending numeric, operational, and relational symbols and understanding numeric morphology and mathematical syntax; older children should be able to read, write, and order fractions.[7] For rapidly developing younger students, tests with equivalent forms afford the opportunity to track their trajectories serially over a short time period.

It is often difficult to detect impairments in these basic skills using tests of more advanced or complex mathematical concepts and procedures. The only two currently available commercial tests that focus exclusively on foundational skills are the Test of Early Mathematics Ability–Third Edition (TEMA-3), which is standardized through age 8 but can be used cautiously to analyze older students' math difficulties, and the Dyscalculia Screener, a British computer-format test for students age 6–14 years.[8] Many researchers have used their own variants and new tests may emerge in the coming years.[9] Other commercial tests include basic items, but usually too few to permit a thorough assessment.

To assess arithmetic fluency, one must rely on speeded tests, which reveal whether the student can retrieve facts automatically by eliminating the option of using alternate, time-consuming strategies, such as adding by counting rather than by automatic recall. Speeded tests are useful because they are sensitive to a wide variety of impairments. By itself, poor performance on a speeded fact test does not provide information about the source of the difficulty, however. Speeded tests of numerical comparison judgments might provide a finer-grained assessment of number sense, but no such test is yet commercially available.[10]

To garner information about a student's problem-solving strategies, untimed tests are essential. One technique for gaining information about how students think through problems is to ask them. Some researchers recommend that the examiner actively probe (e.g., "How did you do it?" "Could you have done it another way?" "How would you explain this to a friend?") or by encouraging students to "think aloud" as they work. (The TEMA-3 actually incorporates such probes.) It is important to note whether the student rigidly draws on one strategy for all problems or flexibly experiments with several different ones. Students sometimes reveal a better understanding of the problem through their hand gestures than they can provide in their verbal explanations; such gestures should be taken into consideration. Finally, the students' errors themselves will also provide information about their misunderstandings and conceptual maturity.[11] These testing approaches require that the examiner know the material; thus, the classroom teacher may be better suited than the psychologist for this aspect of the evaluation.

Early Screening

In the past few years, researchers have begun to devise ways to identify young children who may be at risk for math difficulties, in the hope that very early intervention can prevent problems later.[12] One goal of a screening test is to predict future performance. Longitudinal data on very large groups of students have shown that early competence in number knowledge and number sense was "the most powerful predictor of later learning"—in math *and reading*.[13] Early math skill was a better predictor of later math and reading than were early language and reading skills or even the ability to pay attention. Researchers therefore concluded that individual, expertly administered tests of number knowledge and number sense at the point of school entry "can be a highly reliable way of assessing early skills."[14]

Preschoolers and kindergartners exhibit a wide range of competence levels. Some children, particularly those of low socioeconomic status, enter kindergarten without basic math skills; many of those students fail to progress in them during that year.[15] Studies have found that one can determine from young children's basic number skills whether they are likely to have difficulty with formal math instruction in the early grades. Notably, predictors encompass both number sense skills, such as quantitative comparisons and proximity judgments, and math-language abilities, such as counting and reading numerals—the very skills that we recommend evaluating in anyone complaining of math difficulty. In fact, one ongoing study of 4-year-old children confirmed what we have seen so far in this book: Math skill, even very early math skill, is composed of many partially independent abilities, where "almost no component [is] an absolute prerequisite for any other."[16]

There is not yet a definitive math screening test for use in kindergarten, in part because assessing children at this age of rapid growth is a bit like shooting at a moving target. Many children who enter kindergarten with weak number skills make sufficient gains during the year to take them out of the at-risk category (perhaps justifying screening several times during that year); older kindergartners, particularly those with some basic reading skills, tend to have better basic math skills. Because children often enter kindergarten over a wide age range—sometimes resulting in a gap of as much as 2 years or more between the oldest and youngest students in the class—designing a kindergarten screening instrument is complex.[17]

Researchers are also asking exactly what screening tests will be able to predict and how far out they will be able to do so. So far, one study has successfully predicted low math achievement to third grade, at least under tight diagnostic constraints.[18] Do the kindergarten deficiencies that predict difficulty in third grade also predict difficulty in seventh grade or high school? Other research suggests different cognitive skills may come into play later in the curriculum. In one study, for example, preschool block-building skill strongly predicted math performance in high school, but not at all in elementary school.[19] In the years ahead, tools for early screening will improve as researchers find answers to these questions.

Cognitive Assessment

Myriad skills contribute to learning mathematics. Thus, a comprehensive math assessment will examine analytical problem-solving style and spatial ability, as well as language and basic reading competence. It will also evaluate visual and auditory working memory, the executive functions, and reasoning. Preliminary interviews and data gathering may rule out impairments in one or more of these areas and thus help focus the formal evaluation. Even so, routine screening will ensure that the psychologist does not miss possible contributions to math difficulty. A comprehensive skill survey can also document important cognitive strengths where they exist.

Spatial Skills

Certain spatial skills are important for many mathematical applications and may be associated with number sense. Strengths and weaknesses in spatial skill frequently go unnoticed in school settings, where language and attentional skills are generally sufficient for success in most subjects. Furthermore, tests of nonverbal abilities often focus on iconic functions (e.g., reasoning about shape, color, or pictures) and neglect the spatial factor altogether. Iconic visual skills are important for math, particularly for reading and geometry, but tests of iconic skills should not be confused with those of spatial abilities.

The "cleanest," most useful spatial tests are those least amenable to assistance from verbal input or iconic visual features. Thus tests of two- and three-dimensional mental rotation, which is linked to math ability in some people, are good choices. Among commercially available standardized tests, the Woodcock-Johnson III Tests of Cognitive Abilities Diagnostic Supplement (WJ III SUP)[20] Block Rotation subtest is the closest to the Vandenberg Mental Rotation Test used in much of the research. Although other tests, such as the Wechsler Intelligence Scale for Children–Fourth Edition (WISC-IV)[21] Block Design subtest, require some spatial manipulation, many also rely on visual-motor coordination, iconic visualization, and other strategies; thus, inferences about spatial skill should be made cautiously. Nevertheless, performance on Block Design and on the Position in Space subtest of the Developmental Test of Visual Perception–Second Edition (DTVP-2)[22] have been associated with math ability in some studies.[23]

Other spatial skills are more difficult to evaluate. Given the research, previously cited, demonstrating a relationship between early skills in block building and later math achievement, we wish that a comprehensive, three-dimensional construction test were available; unfortunately, none exists to our knowledge. Nor do most common psychological tests formally examine a child's left-right orientation. Similarly, there is only one available mechanical reasoning test, and it is standardized only for older adolescents and adults.[24] Mathematical evaluations would benefit from availability of tests of all these functions for school-age children.

To assess a student's ability to copy what he or she sees with pencil and paper, a psychologist requires a test of visual-grapho-motor integration, such as the Beery-Buktenica Developmental Test of Visual-Motor Integration (VMI), Fifth Edition[25] or others like it. Such tests capture difficulties in integrating both spatial and iconic perceptual information with grapho-motor behavior—a skill needed for geometry and diagramming.

There are currently no standardized tests to determine visualization problem-solving style. A qualitative performance assessment on some of the complex visual inductive reasoning tests, such as the various matrix reasoning tests, for which analytical rather than holistic imaging skills are required for success, may provide information about students' analytical ability, but will not provide much insight about their preferences. Children are more likely to reveal their problem-solving inclinations on math problems for which either a diagram or a symbolic solution are feasible. Researchers are currently developing a spatial imagery questionnaire that may eventually prove useful in identifying students for whom schematic imagery does not come naturally.[26]

Language and Reading

Receptive and expressive language skills are essential for counting, understanding classroom instruction, explaining one's thinking, and asking questions. These skills include working vocabulary, comprehension, syntax, verbal fluency, object naming, and verbal working memory, as well as underlying phonological skills and articulation. Most psychologists know how to evaluate basic language skills, so we will not belabor that here.

In addition to linguistic competence, reading is also necessary for learning new terminology, facts, and formulas, and for deciphering the compressed and exacting language of math notation, word problems, and textbooks. Because the ability to read math notation and natural-language text are somewhat independent, it is important to evaluate both alphabetic and numeric reading competence. Again, we assume that most psychologists know how to assess the former skill and focus here on the latter.

Skilled mathematical reading, like expert alphabetic reading, requires several fundamental visual abilities, such as scanning the page from left to right and automatically recognizing the symbols. First, a child must be able to discriminate the digits visually. If a student has difficulty matching numerals quickly and accurately, as revealed by a low score on a digit-matching test (e.g., the Woodcock-Johnson III Tests of Cognitive Abilities [WJ III COG][27] Visual Matching speeded subtest), one may suspect weak numeral recognition. The examiner must exercise caution in interpreting such a result, however, because comparably low scores on other *non-symbolic* speeded matching tests (e.g., the WJ III SUP Cross Out subtest) may suggest an ocular-motor rather than (or in addition to) a symbol-recognition problem.

A student also must be able to connect perception to speech. Rapid-naming tests, such as the Rapid Automatized Naming and Rapid Alternating Stimulus (RAN/RAS)[28] test, can detect whether children have learned to read digits and letters fluently. If children name digits as slowly as they do colors or objects, they have not mastered digit naming.[29] Finally, just as children must be able to copy geometric designs, they must also be able to write mathematical symbols fluently. Symbol-copying tasks, such as the WISC-IV Coding speeded subtest, will reveal problems in general symbol copying, although they generally do not require alphanumeric copying.

Executive Functions and Reasoning

The best method for detecting serious executive impairments, such as those that are symptomatic of ADHD, is by taking a good behavioral history. One can quantify behavioral observations by using standardized behavioral rating scales. Two good diagnostic instruments, which both have self- and observer-rated scales, are the Conners 3rd Edition (Conners 3)[30] and the Behavior Rating Inventory of Executive Function (BRIEF).[31] Serious depression can also compromise executive skills. The Children's Depression Inventory (CDI)[32] is a useful screening test for clinical depression in children ages 7 to 17; the Beck Depression Inventory–Second Edition (BDI-II),[33] works well for individuals age 13 and older.

Obtaining ratings from parents and other reliable observers when the examiner suspects ADHD or other executive difficulties is particularly important, because people who have lived with debilitating conditions for a long time often perceive their experiences as normal. Also, some older students seek an ADHD diagnosis to obtain stimulants or academic accommodations and have become savvy enough about ADHD to simulate diagnostic responses on questionnaires. For these students, obtaining behavioral ratings from parents and other observers can help to ensure an objective evaluation and accurate diagnosis.[34]

Some psychologists also employ formal tests that tap aspects of executive functioning. Many of those tests are highly sensitive to executive impairments. For several reasons, however, they must be interpreted with considerable caution. First, the reliability of executive function tests, by their very nature, is difficult to establish. For a test to be reliable, a person must perform comparably on it at two different points in time. A true test of executive problem-solving skill, however, involves a novel task in a novel format—and a test cannot be novel twice. Moreover, a task that is novel for one person may not be so for another.[35] In addition, individuals with attention problems are not always distractible; the problem, more commonly, is that their ability to pay attention is highly variable, so that they may perform poorly on a test at one time and well on the same test at another.

Researchers also often rely on meaningless tasks, such as repeating a verbal number string backward, to assess working memory. Human memory is associative, however, and because even the numbers in simple calculations have meaning (if only their relation to each other), it may be more math relevant to test working memory with meaningful tasks.[36] Alternatively, this sort of test can be used to evaluate an individual's executive ability to impose organization on seemingly random information. Another excellent measure of organizational efficiency is a speeded retrieval fluency test (e.g., "Name as many animals as you can in a minute"), on which one gains a time advantage by thinking of examples in categories (e.g., house pets, zoo creatures, insects). On such a test, however, failure to categorize may signal disorganization in one person, concrete thinking in another, or sluggish verbal processing in a third.

Sustained mental control measures, such as continuous performance tests, depend on various perceptual, motor, and cognitive abilities in addition to executive functions. A student with a deficiency in one or several of these other areas may perform poorly on many of these tests yet still have adequate executive and reasoning skills.[37] For example, for students whose number difficulties stem from perceptual or cognitive impairments, number tasks may serve as poor measures of inattention or other executive and reasoning skills. An analogy would be erroneously determining that students with dyslexia had poor sustained attention solely because they were distractible while reading.[38]

Finally, although students with attention deficits often do poorly on executive function tests, these instruments are not specific enough to diagnose ADHD and may not be adequately sensitive in older individuals with ADHD who have developed compensatory strategies.[39] Moreover, one study demonstrated that older students could simulate ADHD-level performance on several of them, while on others, simulation efforts led to even poorer performance than among individuals with ADHD.[40] A clinician faced with exceptionally poor scores may not recognize malingering. These variable results have compelled Russell Barkley, a prominent ADHD scholar, to claim that long-term behavioral observations by parents and teachers are the best primary diagnostic tools for ADHD.[41] They also provide vital information regarding depression and other clinical conditions.

Despite their drawbacks, however, executive function tests can be a valuable source of information regarding a student's attention, working memory, mental flexibility, and goal-oriented planning ability—all essential to learning and doing mathematics. In fact, performance on most tests depends on adequate attention and mental control, as well as other executive functions; thus, in addition to their intended purposes, the majority of tests in a standard comprehensive psychoeducational battery can provide the alert clinician with a wealth of information about a student's executive functioning. Conversely, any test used to evaluate perception, cognition, or academic skill should be interpreted cautiously if there is other evidence of an executive impairment. For a student with lifelong symptoms of an attention deficit, poor performance on a timed test of visual scanning, for example, may reflect distractibility, impulsiveness, or inflexibility rather than ocular-motor impairment. Administering multiple tests that tap each skill in different ways can serve as a check against confounding factors, much as a good researcher controls for irrelevant variables.

The principle of using multiple, carefully selected tests also applies in evaluating higher-order reasoning. Clinicians often have difficulty locating evidence of good reasoning skill in students who have language or reading difficulties or who are culturally disadvantaged. One must be equally wary, however, of underestimating reasoning skills in students with compromised nonverbal abilities. Given these potential obstacles, the examiner is obliged to actively seek out evidence of a student's best reasoning skills and to interpret test results cautiously. Thus a comprehensive evaluation will routinely include several very different reasoning tests, including visual tasks (e.g., the WJ III COG Concept Formation subtest; the WISC-IV Matrix Reasoning subtest) and verbal tasks (e.g., the WISC-IV Similarities subtest). The WJ III COG Decision Speed subtest is an excellent reasoning test with visual and verbal features. Some tests, such as the Concept Formation and Matrix Reasoning subtests, emphasize logic, whereas others, such as the Similarities and Decision Speed subtests, stress abstract thinking; both sets of skills should be evaluated.

Intelligence

The attentive reader will have noticed that we have not included an IQ test, per se, in our test recommendations, although parts of several such tests are scattered throughout our discussion. As discussed in Chapter 1, the underlying cognitive skills necessary for math are, with a few exceptions, many of the very skills that make up "intelligence." Thus one of the psychologist's main tasks is to find the student's core strengths— particularly in reasoning, but also in spatial, linguistic, and executive skills—because the classroom teacher must know what intellectual resources the student brings to learning. Does that mean for students with few intellectual resources, the prospect for learning math is limited? The extent to which fundamental intellectual faculties can be stretched is a matter of considerable scientific debate.[42]

People learn informally through experience, particularly when engaged in repetitive and salient activities, such as is often found in the work setting. No instruments yet exist that can measure the potential for this type of learning, although it does appear to be independent of abilities measured on standard test batteries.[43] In any case, for individuals with global limitations, common sense would dictate that classroom goals should be geared to the math skills most important to employers. In the past, those requirements were arithmetic operations, decimals, percents, and computing the number of hours worked; an updated survey would provide much-needed guidance.[44] The majority of students, however, will have intellectual peaks and valleys; the psychologist's job is to survey them and know their terrain.

Conclusion

As discussed throughout this book, mathematics requires myriad perceptual, cognitive, executive, and reasoning skills. A comprehensive evaluation of math difficulties will necessarily take these diverse domains into account. Despite the wealth of general scientific evidence, however, no one has yet demonstrated that every individual who has difficulty with math will necessarily reveal impairments on formal psychometric testing, nor that the trouble spots noted on testing necessarily reflect underlying causes of a student's math difficulties. Researchers have made preliminary efforts to answer these questions, but the field remains wide open for future research.

One important factor confounding our diagnostic understanding is time. Difficulties that might seem focal and easily identified when isolated in early childhood may be more global if left undetected until high school or college. By that time, failures in one domain have often stunted growth in others, and a variety of well-developed compensatory strategies cloud the picture. Conversely, younger students do not bring with them the history that can be so helpful in evaluating older individuals. Moreover, students develop at different rates; their development is not always linear or constant and is affected not only by internal forces but also by the impact of different teachers and pedagogy.

Finally, a comprehensive evaluation may identify underlying impairments that have contributed to an individual's math difficulties; however, research on how that knowledge can inform pedagogy is in its infancy. For example, techniques to improve visual-spatial impairments look promising, as do adaptations of well-established reading interventions to enhance mathematical symbol recognition. Until research can provide more definitive pedagogical recommendations, teachers' best recourse will always be simply to know their students well. To that end, comprehensive knowledge of a student's cognitive profile is essential.

Notes

[1] See, e.g., Hale & Fiorello, 2004.

[2] For legal guidelines, see Flanagan, Ortiz, & Alfonso, 2007, pp. 127–141; IDEA 2004 Final Rule, 2006, §300.306(c)(i), §300.307.

[3] See Teisl, Mazzocco, & Myers, 2001.

[4] Barkley, 1997.

[5] See, e.g., Shurtleff, Fay, Abbott, & Berninger, 1988.

[6] Connolly, 2007.

[7] See, e.g., Kalchman, Moss, & Case, 2001.

[8] Ginsburg & Baroody, 2003; Butterworth, 2003.

[9] See, e.g., Deloche et al., 1994; Durand, Hulme, Larkin, & Snowling, 2005; Shalev, Manor, Amir, & Gross-Tsur, 1993; von Aster, 2000.

[10] See Berch, 2005.

[11] See, e.g., Church, 1999; Fleischner, 1994; Ginsburg, Jacobs, & Lopez, 1993.

[12] See, e.g., Chard et al., 2005; Desoete & Grégoire, 2006; Jordan, Kaplan, Olah, & Locuniak, 2006; Mazzocco & Thompson, 2005.

[13] Duncan et al., 2007, p.1443.

[14] Duncan et al., 2007, p. 1444.

[15] Jordan, Kaplan, Locuniak, & Ramineni, 2007.

[16] Dowker, 2005, p. 330. See also Gersten, Jordan, & Flojo, 2005; Mazzocco & Thompson, 2005.

[17] Jordan et al., 2007; Mazzocco, 2005; Teisl et al., 2001.

[18]Mazzocco & Thompson, 2005.
[19]Mazzocco, 2005; Wolfgang, Stannard, & Jones, 2001.
[20]Schrank, Mather, McGrew, & Woodcock, 2003.
[21]Wechsler, 2003.
[22]Hammill, Pearson, & Voress, 1993.
[23]See, e.g., Hegarty & Kozhevnikov, 1999; Mazzocco & Myers, 2003.
[24]Bennett, Seashore, & Wesman, 1990.
[25]Beery & Beery, 2004.
[26]Blajenkova, Kozhevnikov, & Motes, 2006.
[27]Woodcock, McGrew, & Mather, 2001.
[28]Wolf & Denckla, 2005.
[29]See Wolf & Denckla, 2005.
[30]Conners, 2008.
[31]Gioia, Isquith, Guy, & Kenworthy, 2000; Guy, Isquith, & Gioia, 2004.
[32]Kovacs, 1992.
[33]Beck, Steer, & Brown, 1996.
[34]Booksh, 2005.
[35]Phillips, 1997.
[36]See, e.g., Biederman et al., 2006; Passolunghi & Siegel, 2004.
[37]Phillips, 1997.
[38]Advokat, Martino, Hill, & Gouvier, 2007.
[39]Barkley, 1997; Kaplan & Stevens, 2002.
[40]Booksh, 2005.
[41]Barkley, 1997, pp. 331–337.
[42]See, e.g., Detterman, 1993; Shalev, Manor, & Gross-Tsur, 2005; Stevens et al., 2003.
[43]Mackintosh, 1998.
[44]Algozzine, O'Shea, Crews, & Stoddard, 1987.

Chapter 11

Teaching

M athematical pedagogy informed by cognitive research is in its infancy. Although most students are remarkably adaptable and do well with common teaching practices, many students do not. For the myriad reasons outlined previously, these students miss out on some or all of the mathematics curriculum, with enormous cost to their education, self-confidence, and future job prospects, as well as to the nation's standing in the global economy. These difficulties, as they present themselves daily in the mathematics classroom, are frustrating for students, teachers, parents, school administrators, curriculum developers, and textbook publishers. As some of our colleagues have wistfully noted, there is no Orton-Gillingham teaching method for math disabilities, referring to a scientifically based pedagogical approach to dyslexia.[1]

This chapter summarizes the educational implications discussed in the research chapters. We then briefly discuss a number of issues, some of them controversial, affecting teachers: student assessment, the proliferation of standards and options, teacher development, and what to look for in educational research. The goal of this chapter, in setting out the issues, is not to resolve them—the educational community is still a long way from that—but to lay the groundwork for future discussions and research.

Summary of Research Implications for Pedagogy

Research suggests that students require three general sets of skills to learn mathematics. The first is number sense, a mental map of the quantitative terrain and a skill closely related to spatial sense. The second is mastery of the written symbolic conventions and how they express quantitative and other mathematical relationships. Third, students require certain general functions, including executive and reasoning skills, that are not specific to mathematics but are necessary for solving problems.

Number Sense

As discussed in Chapter 2, one thinks of numbers and other ordered concepts as if along a line, and researchers have understood students' ability to place numbers along a physical line as reflecting their mental representation of number. Studies of young children at risk for math learning difficulties have shown that number-line activities (e.g., linear numerical board games) produce clear, salutary, and lasting effects on a wide range of number skills. This is a very promising line of research. Educators should look for future number-line studies in relation to older children and to more advanced number concepts, such as fractions. Broadly speaking, the research points to the importance of fostering a sense of numerical relationships (e.g., 12 is more than 10, a lot more than 2, and less than 15), as well as a sense of reasonableness in problem solving. This comfort with numbers—this number sense—enhances all quantitative manipulations and is the bedrock mathematical skill. Future research will indicate whether number-line activities are useful at all levels of mathematical education.

Spatial skill and a schematic way of thinking about problems are important not just for quantitative understanding, math-fact mastery, and computations, but for many general mathematical concepts, including geometry, logic, complex functions, and their applications in a broad range of fields. Preliminary research suggests that spatial skill can be improved through a wide variety of activities, such as playing Tetris, studying music, or constructing clothing patterns—although how long the improvements last and whether they benefit students with the most profound impairments in spatial skill, for example, remain unclear. The questions of whether such activities foster the habit of looking at problems schematically or improve number sense and general math achievement also remain unresolved. Additionally, researchers have not yet found a way to treat impairments in integrating visual-spatial and grapho-motor skills (copying figures with pencil and paper), an ability that is fundamental to many mathematical tasks, including geometry.

The Language of Mathematics

Unlike the approximate and relative notions of number that children grasp intuitively when they are young, the exact and absolute ideas of number required by daily life are tied to culture and must be learned through hard work. In particular, exact mathematics relies on oral and written language, beginning with counting. Mathematical language holds certain features in common with natural language, such as the underlying phonological, articulatory, oculomotor, visual, grapho-motor, and memory requirements. Other facets, however, are specific to conventional mathematics, which has a vocabulary, grammar, and written notation of its own. Mastering mathematical language is a complex task that involves integrating these various components, much as in learning natural language and alphabetic reading. Brain studies confirm a very imperfect cognitive overlap between natural and mathematical language; disruptions in one are often, but not always, tied to disruptions in the other.

Most children begin learning the spoken aspects of mathematical language, such as counting, spatial vocabulary, and number-related grammar, alongside natural language well before they start school through informal instruction accompanying routine daily activities. Similarly, they learn symbolic mathematical notation in tandem with alphabetic notation in the early school years. For many children, the most difficult part of learning formal mathematical language is connecting it to their informal quantitative intuition. Although much more work remains to be done, research suggests that teaching counting and basic arithmetic notation and concepts in the context of number-line activities may foster this important link. Whether number-line instruction would also help students link their intuitive sense of proportion with the notoriously difficult and unintuitive fractional notation awaits future research.

For retention of information, such as arithmetic facts, both strategy instruction and drill are essential, although which facts need to be drilled and in what format remain open questions. Format may be a particularly important question for children with cognitive impairments affecting specific learning modalities. Spaced practice is more effective than cramming, and repeated quizzing works better than prolonged study. Early finger use helps children gain experience with correct results, which boosts mastery. In general, corrective feedback prevents students from "learning" their errors. As with learning facts, algorithmic competence requires prolonged and focused practice and is greatly enhanced by systematic instruction that increases a student's strategy options. In this manner, factual, procedural, and conceptual instruction and mastery support each other. The long-term impact of calculator use is still not well understood, although it may affect fact- and procedural-learning differently.

The strong linguistic component of conventional mathematics, while challenging for many students under ordinary conditions, often spells particular trouble for students with language or reading disorders and for students from linguistically disadvantaged homes. To our knowledge, no research exists on applying or accommodating the well-established multisensory speech, language, or reading and writing therapies to mathematics education. For teaching comprehension of math vocabulary, explicit and systematic instruction appears promising. Work in this area is long overdue. In this era of globalization and geographic mobility, teachers need to learn more about the special needs of students who begin their mathematical education in one language and complete it in another. Some mathematical knowledge, such as terminology or multiplication facts, depends significantly on the particular language in which a student learns it and may need to be relearned in the second language.

Solving Problems

Learning mathematics also entails several mental functions that are not specific to math, but that are necessary for problem-solving and general adaptation. These capabilities and the front portion of the brain, upon which they largely depend, mature gradually with learning and experience, often well into adolescence or early adulthood.

One such set of skills, known as the executive functions, includes attention and mental control, working memory, planning and sequencing, self-monitoring, and mental flexibility—all key components of math learning. Most teachers know instinctively that they can expect more of older students than younger ones, who require more guidance, and adjust their pedagogy accordingly. But anxious and preoccupied students and those with specific neuropsychiatric disorders, such as ADHD, affecting executive functions require special consideration. Although there is no cure for executive impairments, research indicates that classroom techniques and accommodations can often partially ameliorate their negative effects. Predictable structure, explicit instruction, immediate feedback, judicious rewards, and "think-aloud" exercises have all been shown to help. Reducing classroom distractions, breaking up lessons to limit the load on sustained attention and working memory, distributing information across verbal and visual modalities, and providing regular outlets for excess energy have also been proven beneficial.

For children with ADHD and other neuropsychiatric disorders, research has demonstrated that medication can produce short-term performance improvement, but whether it can provide long-term gains in math achievement remains unresolved. While children with executive impairments have trouble with facts and procedures, they can often get the big ideas and should be provided with plenty of opportunity to explore those. More research on these techniques as they apply to the math classroom in particular is warranted. Methods aimed at ameliorating math anxiety, a source of disruption whose effects are limited to mathematics performance, have not met with reliable success and none have improved students' math skills.

The second set of problem-solving skills includes abstract thinking and logical reasoning, the so-called higher-level cognitive abilities most closely linked to intelligence. These skills lie at the heart of all mathematics (indeed, all learning and intellectual activity) and relate to the sticky question of transfer, or the ability to generalize specific learning to broadly serviceable problem-solving skill. Although logic, for example, was a key component of classical education, few studies have addressed how to teach it. Whether and to what extent transfer can be taught is the subject of spirited scholarly debate.

Optimistic psychologists and educators highlight several pedagogical techniques aimed at fostering general problem solving. These approaches include explicitly pointing out abstract links and concrete distracters, analyzing and anticipating faulty assumptions, warning students about common pitfalls, and focusing on the problem set-up apart from its solution. They also include introducing new concepts with worked examples, pointing out the common mathematical structure in diverse problems, asking students to discover the solution-relevant information in real-life settings, and encouraging students to generate their own problems, data, and formulas. A few of these methods have garnered some research support for their efficacy in the math classroom, but many questions remain unanswered and considerably more research is warranted.

Understanding Your Student

Each child has a unique cognitive profile. No matter what instructional materials a district or school has chosen, the teacher on the front line must adapt it to the needs of specific students. Even students who perform adequately may learn in diverse ways. For struggling students, as the vignettes throughout *Number Sense and Number Nonsense* demonstrate, determining the underlying issue(s) for each student is a critical first step in developing an intervention plan, formal or otherwise. Chapter 10 focused on the importance of, and options for, obtaining a portrait of such a student's reasoning and perceptual, cognitive, and executive functioning. On rare occasions, this information will be found in a student's folder in the school office. In addition to reading such reports, the teacher also must be aware of the student's social and emotional state. Working with the school or private psychologist to determine a student's reasoning ability and spatial, linguistic, and executive skills is ideal, but not always feasible.

Teachers taught to observe their students carefully, however, can obtain much important and relevant information for themselves. In fact, standards documents, such as *Principles and Standards for School Mathematics* of the National Council of Teachers of Mathematics (NCTM), require that teachers have the knowledge and ability to analyze students' thinking. Research aimed at understanding and measuring the methods that teachers naturally use to do so, however, is still in an early phase in this country. Not much is yet known about how well teachers' understanding of students' thinking predicts student outcome, although recent studies have discovered that teachers' ratings of young students' *behavior* is related to math achievement.[2]

One method that teachers have at their disposal to track their students is formative assessment, a systematic process designed to gather evidence at brief intervals (every week or two) about a student's progress for the purpose of informing instruction and providing student feedback. Ideally, teachers use a range of short instruments, from well-researched standardized and curriculum-based tests to informal "think-alouds," that assess factual, procedural, conceptual or problem-solving mastery as needed. When performed as a regular part of classroom practice, the instruments allow teachers to detect error and misunderstanding early and to correct them—perhaps by trying a different approach—before these problems can create serious mischief (and thus avoid the time commitment and frustration of having to reteach the material later). For this reason, formative assessment has been deemed superior to summative assessment, which typically occurs at the end of a unit or course and gauges student progress toward course or grade-level benchmarks. While the latter is like a series of snapshots, the former resembles a moving picture.[3]

Because formative assessments occur frequently and focus on the individual student, they can serve as a window—a microscope, even—into how the student thinks and grasps new ideas. For the student who did not understand a concept last week but does today, the teacher can ask, What happened? What clicked? What aspect of instruction or that particular student's classroom experience allowed the idea to bloom at this time? What intellectual or psychological change—what educators call the *zone of proximal development*—prepared the way for understanding? Reflections and observations such as these give the teacher important clues about how a student learns, which in turn can improve pedagogy.

In the current climate of antipathy to testing, many educators still tend to regard formative assessment as something external to, or even incompatible with, daily practice, yet one more chore for which they have no time. Teachers are all too aware, however, that the annual state tests provide too little information that arrives too late to contribute to planning for the current instructional year. Although many districts have developed interim assessments to solve that problem, even those tests are administered too infrequently and cover too much material to be of much instructional use.[4] Teachers would do well to recognize that they (the teachers themselves) are the very best sources of information about their own students. Although psychologists and other specialists can add valuable insight, no one else sees the student as frequently or knows his or her thinking as intimately as the front-line teacher.

Proliferation of Standards and Options

The No Child Left Behind (NCLB) Act of 2001 (PL 107-110) requires that states adopt challenging academic standards that specify what students must know and be able to do, as well as encourage teaching of advanced skills.[5] According to one study, 38 states have revised their mathematics standards since 2002, although states differ in whether the standards are models for local districts or are mandatory requirements for each student to master. The specificity and language of learning goals and the grade placement of specific math topics vary significantly across states, perpetuating the publication of math textbooks, designed to be marketed in multiple states, with repetitive, misaligned, or superficial coverage.[6]

Attempts at Coherence

The NCTM released a series of documents designed to furnish direction and achieve coherence in mathematics instruction. These documents provide a history of reform efforts.[7] In 1980, *An Agenda for Action* made 10 key recommendations including a new emphasis on the need for students to learn how to solve problems.[8] Nine years later, the *Curriculum and Evaluation Standards for School Mathematics* set out to turn the 10 recommendations into a vision for teaching mathematics in grade bands from prekindergarten to grade 12.[9] This document and two later documents on teaching and assessment were further expanded in 2000 with the *Principles and Standards for School Mathematics*, which added five new process standards—communication, reasoning, representation, connections, and problem solving.[10] In theory, these process standards should pervade mathematics instruction, regardless of curriculum.

More recently, NCTM has developed *Curriculum Focal Points: A Quest for Coherence*, with the self-proclaimed goal of promoting collaborative discussions about areas emphasized in each grade from prekindergarten through eighth grade.[11] This latest document replaces the long lists of specific goals that a child should have met at any given grade with a description of the most significant mathematical concepts and skills for each grade level.

In 2008, the National Mathematics Advisory Panel (NMAP) added another voice to the already confusing mix. The NMAP reviewed myriad studies, but in considering only the most rigorously designed and executed ones, found few that supported pedagogical recommendations.[12] Not surprisingly, they found that there is no one ideal approach that is appropriate for all students. The students' background, abilities, and effort; the teachers' background and strengths; and the instructional context, approach, and curriculum should all factor into pedagogy. The NMAP cautioned that much rigorous investigation will still be required to enable researchers to predict which instructional materials and methods are effective for which particular teachers, students, topics, and learning conditions. The report recommended[13]

- Formative assessment of what students understand and are able to do mathematically on a regular basis

- Design of instruction based upon this assessment and responding to individual students' strengths and weaknesses *and* the research when available

- Use of instructional approaches and tools suited to the mathematical goals of the classroom, realizing that a mix of teacher-directed and student-centered instruction will be needed

- Teacher openness to participation in research studies to answer the unresolved questions in the educational field

Given this profusion of curricular guidelines and requirements, textbook misalignments, and pedagogical uncertainties, teachers and district curriculum directors are often still at sea when deciding what skills and concepts to teach, as well as when and how to teach them. For this reason, among others, cognitive and developmental research will continue to play a key role in curriculum development.

Curricular and Pedagogical Options

Much remains unknown about curricular and pedagogical effectiveness and how to choose among the available options. How to implement curriculum and process standards is as hotly debated as the standards themselves. We briefly review three sets of options.

One important issue is the structure of the curriculum. Most adults remember studying topics in systematic, linear order, mastering one topic before moving on to the next. More recently, publishers have introduced a curriculum that spirals, wherein students are briefly introduced to concepts that then reappear later in the year and in subsequent grades. This type of curriculum can be appropriate for many students, but has several disadvantages for those who struggle, including limited time to understand a concept, insufficient practice opportunities, and consequent gaps in their understanding when the concept is expanded at a later time.[14] Parents and teachers look forward to research that resolves this issue.

One of the most controversial topics in math education is the relative effectiveness of student-directed, or cooperative, learning versus teacher-directed instruction. Cooperative learning groups have been popular since the late 1970s, particularly for remedial tutoring and collaborative problem solving.[15] When students teach each other, however, they risk missing key concepts or developing serious misconceptions. Not surprisingly, the NMAP found in their research review that effective instruction is subtler and more nuanced than advocates of one position or the other might argue. Despite the controversial nature of the subject, the NMAP identified only eight technically adequate studies that directly compared the two instructional styles, and failed to find sufficiently compelling evidence to recommend either practice in isolation.[16]

The findings are somewhat clearer regarding effective pedagogy for low-achieving students. The NMAP reviewed 26 high-quality studies pertaining to teaching mathematics to struggling students and concluded that explicit, teacher-directed instruction is consistently and significantly superior to other methods. Key components of this approach include providing carefully sequenced problems, clear models of proficient performance, frequent opportunities to verbalize solution strategies, and adequate practice and review.[17] Although teacher-directed instruction should not be the only pedagogical approach, teachers must ensure that students possess foundational skills, conceptual knowledge, and computational proficiency. This is a chief criticism of many curricula that expect students to reach procedural fluency without direct teaching.

Another controversial topic in mathematics education is the use of real-world problems to engage students and teach problem solving. Called *cognitively guided instruction* in some research, real-world problems are used to engage students in the five NCTM process standards.[18] The definition of *real-world*, however, depends upon the author; moreover, what is real to one student is not necessarily so to another. The NMAP concluded that research findings were mixed, with real-world problem solving leading to gains in some areas with some students. (As discussed in Chapter 9, scholars similarly debate the relative pedagogical value of concrete versus abstract examples.) When subsequent problems were similar, students tended to do better; however, performance on computations, simple word problems, and equation solving did not improve.[19]

Teacher Considerations

Guidelines, standards, curriculum, and pedagogy are only some of the terms in the education equation, however. Another key element is what the teacher brings to the classroom. This section discusses teacher training and examines some cross-cultural differences in educators' philosophy, math training, and approach to teaching.

Teacher Preparation

Teacher training experts recommend that elementary school teachers, and particularly those who teach students with special needs, should have a deep understanding of the following areas. [20]

- *Content*—the mathematics one teaches and how it relates to students' earlier and later instruction

- *Methods*—effective teaching techniques for one's particular students and math topics

- *Mathematical psychology*—the development of children's mathematical thinking and knowledge

Unfortunately, not all teacher training programs adequately prepare new teachers in each of these areas. Studies have found that elementary education majors have fairly high levels of math anxiety, particularly those in early and special education.[21] (Indeed, teachers' own early schooling, combined with attitudes and beliefs about math developed during college and teacher training, can have a profound effect on their later interactions with students.[22]) However, young children are capable of understanding more about number and arithmetic than previously thought possible by parents, teachers, and administrators, placing an extra burden on mathematically insecure teachers. Thus teacher preparation programs are often called upon to fill the gaps in a prospective educator's own schooling—a goal they may not be able to meet. Special education teachers, for example, are usually expected to understand K–12 mathematics curricula, but often receive little relevant preservice instruction.[23]

Educators must be solidly grounded in mathematical psychology to develop the well-planned formal and informal experiences so crucial to effective learning.[24] For a curriculum to be taught coherently across the grades, with a focus on key mathematical concepts (and not a mere collection of activities), teachers must be prepared to plan well-organized and sequenced lessons.[25] To this end, teachers must be knowledgeable about factors both internal (the child's developmental readiness, cognitive strengths and weaknesses, and interest) and external (the subject matter and teaching methods) affecting learning experiences.[26] American elementary teachers are typically expected to be jacks of all trades, however, and have limited time and resources for tending to individual students or developing expertise in any one subject area.

U.S. versus Asian Mathematics Education

East Asian students' superiority in mathematics, from the early years through adolescence, is well established and has drawn considerable interest from American educators.[27] What are the differences in mathematics education, and what can one learn from our East Asian colleagues?

Math education in China, Japan, Korea, and Taiwan is founded on an in-depth knowledge of mathematics and stresses depth over breadth. One reflection of this is that Chinese teachers are specialists in their subject area, even in elementary school; those who teach math teach nothing else. By contrast, U.S. elementary school teachers are typically expected to teach reading, language arts, social studies, and science, as well as math—effectively precluding specialization in any one field. American teachers also have substantially less planning time to devote to daily preparation, reflection, and interaction with colleagues than do their Chinese counterparts.[28] In fact, the two cultures have different views about the importance of nurturing teachers' intellectual and professional needs and development. For example, Chinese and Japanese teachers communicate regularly with colleagues both in and out of school, in written publications, lesson study, and scholarly discussions—activities facilitated by having their desks in a communal teachers' room rather than in separate classrooms, as their more isolated American colleagues do.[29] With more opportunity to plan, reflect, and communicate, American teachers might be better able to tend to the diverse needs of their individual students and to hone their math skills.

In addition, East Asian teachers grow up with the idea, passed down culturally, that effort is more important to learning than ability. Few people would dispute that it is quite common and acceptable in the United States to hear someone say that he or she is "not good at math." Such a view suggests that one believes ability is fixed and uncontrollable, which limits motivation to work hard and become better. By contrast, East Asian teachers emphasize effort over ability, leading students to sustain their belief in the possibility of success and persist in the face of difficulty.[30] Interestingly, we have not seen any studies describing math learning disabilities among students in East Asian countries. The reason for this dearth in the literature is unclear, although it would probably be a mistake to assume that such disabilities do not exist (developmental dyslexia is known in China).[31]

Studies have found that, as a reflection of their mathematical expertise and interest, East Asian teachers emphasize problem-solving methods and their rationales, so that students can apply them broadly; to this end, they spend most of their time discussing a single problem in depth and from multiple viewpoints. American teachers, by contrast, typically focus on procedures, followed by many short practice problems, stressing a single, correct solution method; U.S. students and teachers alike in these studies were often unconcerned if pupils could not explain how or why the method worked.[32] Interestingly, a study of U.S. state curriculum standards found that of the grade-level indicators studied, 153 called for students to express or describe rules, while only 48 called for understanding and application, 10 for analysis, and 5 for explanation and justification.[33] In fact, Asian math curricula in general have been found to be both deeper and more coherent than those used in the United States.

An emphasis on depth over breadth has implications for high-stakes testing. In China, one long-term goal of math education is to prepare students for their university entrance examinations; assessments from the earliest grades through high school are modeled after them. These high-stakes tests, which are written by mathematics specialists, expect students to demonstrate understanding through proofs, constructed responses, and problem solving. By contrast, U.S. teachers have criticized demands to "teach to the test" as unrelated to classroom goals—and in the U.S., most high-stakes tests assess only basic proficiency. Although there may be many factors contributing to teachers' discontent on this matter, including the ever-shifting nature of American standards, one chief source of teachers' frustration is the sheer scope of material.[34]

The NMAP has strongly recommended reducing the number of topics covered in American math classrooms and textbooks, while covering the retained topics in greater depth—a view shared by other researchers.[35] This approach will be possible, however, only to the extent that teachers themselves have an in-depth understanding of the concepts, which is a matter that goes back to their own schooling and professional training.

What to Look for in Educational Research

For a classroom teacher or curriculum director bombarded by educational research reports, textbook sales pitches, or classroom product marketing, it is difficult to know what to look for or what constitutes quality research. Furthermore, one must discern what findings are applicable to a wide range of classrooms versus those results that apply only narrowly. One educator, Stephen Davis, argues that most school administrators, teachers, and college professors are not sophisticated consumers of educational research, and that most researchers have lost touch with the complexity of the human interactions in schools. He warns that not all research is good research and encourages educators to rely on refereed research publications, learn the fundamentals of research design from credentialed experts, and read multiple articles on each topic. Educators should compare and contrast methods, look for common threads and discrepant information, and check for supporting research citations. Often even the best research does not make it into the hands of the classroom teacher because important discoveries travel through separate professional pipelines. Numerous variations in national, state, and local policies; in local political climates; in management philosophies; in student characteristics; in teachers' education, skill, and philosophy; and in available resources all influence how research findings are actually applied in a given classroom.[36]

To help teachers navigate the maze of educational research, this section briefly reviews several general research designs. This list is not comprehensive, but should provide a basic orientation to the methodological issues.

Qualitative versus Quantitative Research

Since the late 1970s, scholars have argued over the relative superiority of quantitative versus qualitative research. Some researchers hold that quantitative research is more rigorous and less susceptible to bias than qualitative research. Others, by contrast, posit that while quantitative research allows for generalization to large populations of students, it fails to capture the human interactions, contexts, and depths of understanding in schools—features better captured by a qualitative study design, in their view.[37]

The NCLB Act came down strongly in favor of quantitative research. Its standards for deciding the effectiveness of "scientifically based" pedagogy include the following categories.[38]

- *Strong evidence* of effectiveness can be demonstrated using randomized controlled trials showing effectiveness in two or more typical school settings, including a setting similar to that of the school or classroom to which one hopes to generalize the results.

- *Possible evidence* of effectiveness may be considered when a randomized controlled trial is inadequately executed or has an inadequate sample size. It may also be considered with comparison group studies if the intervention and comparison groups are closely matched.

- *Studies* that do not meet the foregoing criteria do not qualify as scientifically based.

We discuss each category in turn.

Randomized Controlled Studies

The best research on new pedagogical strategies, curricula, and classroom materials has two key features: it compares the intervention of interest to some other method (usually the standard classroom practice) and is free of bias. An intervention is typically tested on one group of students, and its effectiveness is compared to that of standard practice on another group of students (referred to as the control condition). Ideally, the intervention participants should experience all of the conditions experienced by the control participants except for the particular intervention of interest. That way, the results will say something specific and useful about the new practice. Comparing a new intervention's effectiveness to that of the standard method is of enormous practical importance, of course, in that the results will then indicate whether it makes sense to revise classroom practice.

Good research is also free of bias. One method to ensure this is to assign students randomly to the two conditions. In this way, there are no systemic student group differences that could affect the outcome. For example, if the intervention trial participants were stronger math students, it might lead to spurious positive results. If the two groups resemble each other, however, then any difference in the outcome can be attributed to fundamental differences in the teaching methods. (One possible reason for the relative dearth of good education studies is that parents and teachers sometimes balk at assigning students to one condition or the other—depending, of course, on how they feel about the intervention.[39]) Other sources of bias, such as the exclusive use of subjective outcome measures or the researchers' having a financial interest in the results, should also be avoided.

Besides being bias free and controlled, good studies will also produce results that can be interpreted and generalized. To this end, they require a large enough number of participants (e.g., students, classes, schools); statistically speaking, the more participants involved in the study, the more confident one can be in the findings. It is also important that the research specify the nature of a study's site (e.g., inner city, wealthy suburb) and participants (e.g., do they comprise the full ability range or just one end of it), so that readers can determine whether the results apply to their own students and circumstances. Likewise, the study should detail the specific procedures used in the intervention, allowing readers to faithfully duplicate them in their own classrooms. The measures used to determine the intervention's effectiveness should be reliable, objective, and meaningful, so that readers can know what results they may expect from the intervention when they use it. Researchers should report the study's outcomes in plain English, for obvious reasons, and include negative as well as positive findings, because it is important to know when the intervention of interest is not better than the alternative. Finally, many academic studies take cross-sectional "snapshots," but the most useful education studies are those that collect long-term data to determine whether the improvements can be sustained over time.[40]

Possible Evidence of Effectiveness

Studies that fall short of these basic research standards may, nevertheless, provide some support for an intervention. For example, some studies investigate only one teaching site, fail to randomize properly, or have other flaws, such as high attrition. Other studies demonstrate an intervention's effectiveness in a laboratory but do not test the method in a more realistic classroom setting. Studies such as these should be interpreted cautiously by the reader; they should also be replicated using a tighter or more comprehensive research design.

Some studies do not randomize students into groups, but rather compare existing groups that may or may not be comparable. In these sorts of studies, one can draw conclusions about the intervention's effectiveness only if the groups are closely matched in terms of academic achievement, standardized test scores, demographics, and other characteristics prior to the intervention.

Nonqualifying Research

One is urged to be wary of often popular and seemingly sensible, but inadequate, research designs. Pre-post studies, which simply compare student performance before and after an intervention, are flawed because they do not account for progress that the student might have made anyway without the intervention. Another research genre widely available to classroom teachers is known as action, or teacher, research. Situated in a particular teaching context, and thus precluding generalizability, action research is based on teachers' self-evaluations. Action research studies can prompt teachers to rethink and change their own teaching practices and are, therefore, appealing, but their design is seriously flawed. Inherently troubling issues include how to measure success and how to avoid bias when the researcher is the practitioner.[41] Nonrandomized studies in which the groups are not well matched also do not meet NCLB standards for scientifically based research. Research sponsored by groups or individuals with political agendas or financial conflicts of interest represent the potential for bias in research and marketing, and should raise a red flag for consumers. In fact, any study that does not meet strict research criteria, such as having too few participants, inconsistent procedures, or researcher bias, as well as those based on anecdotal evidence, cannot be used to support an educational practice.[42] Clearly, this includes qualitative research.

Application of Neuroscience Research in the Classroom

The NMAP cautions that "attempts to connect research in the brain sciences to classroom teaching and student learning in mathematics are premature. Instructional programs in mathematics that claim to be based on brain sciences research remain to be validated."[43] Therefore, teachers and curriculum directors should routinely ask questions and request research reports to determine whether an educational theory or product purporting to be "brain-based" has been tested according to the standards discussed previously, thereby putting the burden of proof on the salesperson pitching the product or program. Evidence, as always, should be statistically convincing. One looks forward to the day when brain research can inform pedagogy in the trenches. In the meantime, educators should interpret all "brain-based" claims with skepticism.

Conclusion

Successful mathematics teaching depends on many vital factors. Some, such as community support, local politics, and demographics, are usually not under educators' control, but many other factors are. These include the curriculum, teachers' mathematical expertise, teachers' knowledge of their students, and professional development.

All evidence indicates that math curriculum is in chaos in the United States. As we have seen, learning mathematics is a highly complex cognitive undertaking, even at the most basic level. When teachers are forced to rush students through the fundamentals in order to cover myriad and chaotically organized topics, everyone loses. The NMAP's recommendation that curriculum developers and textbook publishers streamline their products and devote more time and space to the basics is an excellent one.

Curriculum simplification and emphasis on in-depth knowledge of the basics will require that teachers be prepared accordingly. Number sense—comfort with numbers and their relationships, confidence in manipulating them, and the ability to look at problems from many different angles—is the most fundamental quantitative skill. One should expect all mathematics teachers to have good number sense, but it is especially crucial for those at the earliest grades, where the foundations are built and student habits are formed. Teachers with number sense will see opportunities to teach about number in all kinds of otherwise mundane classroom activities; one would hope that their natural curiosity and fearlessness about number would rub off on their young and impressionable students. Unfortunately, although researchers have begun to figure out how to teach number sense to very young children, they have not yet found a way to teach, enhance, or restore it with older students, such as teachers in training.

Successful math teaching also depends on teachers' knowledge of their students. Our readers have probably all had at least one math teacher for whom mathematics came so effortlessly that he or she could not understand why anyone might fail to understand it. Explaining a concept to a struggling student requires that the teacher be able to comprehend, or even anticipate, his or her misconceptions. To do that, in turn, requires understanding how the particular student thinks. This is necessary for all educators, not just special education teachers. For this reason, professional training should include coursework in cognition and cognitive development, particularly as it relates to quantitative thinking. Although neuroscience has not yet produced definitive pedagogical guidelines, it will play a major role in future advances; thus prospective teachers should have a working knowledge of it, as well. Knowing the rudiments of the relevant psychology, furthermore, will enable teachers to know when to seek consultation with other professionals, such as school psychologists and speech, language, and reading specialists, and to know what kind of help to expect from them. As we have seen, teachers play a central role in these collaborative efforts.

Finally, professional development—preservice and inservice—should include basic instruction in research methods. Most teachers will not have the opportunity or interest to engage in research themselves, but all should be expected to read and judge the work of others. To this end, teachers also require easy access to published studies. Although most research is available online, it is often written in the technical language of diverse disciplines and appears in myriad academic journals accessible only through personal or library subscription. Knowledgeable reviews of current research in professional newsletters would be helpful, but teachers need to be able to read the original reports for themselves. Conversely, teachers' viewpoints and input should play an important role in pedagogical research because they know their students and the classroom issues best.

Notes

[1]See, e.g., Clark & Uhry, 1995; Meyler, Keller, Cherkassky, Gabrieli, & Just, 2008.

[2]Allsopp, Kyger, & Lovin, 2007; Fuchs, Fuchs, Compton, et al., 2006; Hill, Ball, & Schilling, 2008; National Council of Teachers of Mathematics, 2000.

[3]See Bloom, Hastings, & Madaus, 1971; Duffy & Kear, 2007; Fisher & Frey, 2007; Gersten et al., 2008; Guskey, 2008; Heritage, 2007.

[4]Heritage, 2007.

[5]No Child Left Behind Act of 2001 (PL 107-110). See also http://www.ed.gov/nclb/landing .jhtml

[6]Reys & Lappan, 2007.

[7]Allsopp et al., 2007; National Council of Teachers of Mathematics, 1989, 1991, 1995, 2000.

[8]National Council of Teachers of Mathematics, 1980.

[9]National Council of Teachers of Mathematics, 1989.

[10]National Council of Teachers of Mathematics, 1991, 1995, 2000.

[11]National Council of Teachers of Mathematics, 2006.

[12]National Mathematics Advisory Panel, 2008.

[13]Gersten et al., 2008, p. 209.

[14]Allsopp et al., 2007.

[15]Gersten et al., 2008.

[16]Gersten et al., 2008.

[17]Gersten et al., 2008.

[18]Carpenter, Fennema, Franke, Levi, & Empson, 1999.

[19]Gersten et al., 2008.

[20]See, e.g., Baroody, 1987.

[21]Hembree, 1990; Baroody, 2004; Ashcraft, Kirk, & Hopko, 1998; McLeod, 1992.

[22]Kessel & Ma, 2001.

[23]Allsopp et al., 2007; Chinn & Ashcroft, 2007.

[24]National Council of Teachers of Mathematics, 2000; Baroody, 2004.

[25]National Council of Teachers of Mathematics, 2000, p. 14.

[26]Baroody, 1987; National Council of Teachers of Mathematics, 2000.

[27]See, e.g., Gonzales et al., 2004, Tables C3, C4; Organisation for Economic Co-Operation and Development, 2007, Table 6.2c; Siegler & Mu, 2008.

[28]Ma, 1999.

[29]Stigler & Perry, 1998.

[30]Newton, 2007; Schmidt, Houang, & Cogan, 2002; Hiebert et al., 2003.

[31]See, e.g., Yin & Weekes, 2003.

[32]Kessel & Ma, 2001; Stevenson & Stigler, 1992; Stigler & Hiebert, 1999.

[33]Reys & Lappan, 2007.

[34]Schmidt, 2001.

[35]Gersten et al., 2008; Reys & Lappan, 2007; Newton, 2007.

[36]Davis, 2007; Willis, 2007.

[37]Kelly & Lesh, 2000; Romberg & Collins, 2000.

[38]U.S. Department of Education, National Center for Education Evaluation and Regional Assistance, 2003.

[39]See Steen, 2007.

[40]Davis, 2007.

[41]Doerr & Tinto, 2000; Feldman & Minstrell, 2000.

[42]U.S. Department of Education, National Center for Education Evaluation and Regional Assistance, 2003.

[43]Geary et al., 2008, p. 163.

Bibliography

Abedi, J., Hofstetter, C., Baker, E., & Lord, C. (2001). *NAEP math performance and test accommodations: Interactions with student language background* (CSE Technical Report 536). Retrieved on March 6, 2007, from http://www.cse.ucla.edu/Reports/newTR536.pdf

Abedi, J., & Lord, C. (2001). The language factor in mathematics tests. *Applied Measurement in Education, 14,* 219–234.

Abikoff, H., Courtney, M., Szeibel, P., & Koplewicz, H. (1996). The effects of auditory stimulation on the arithmetic performance of children with ADHD and non-disabled children. *Journal of Learning Disabilities, 29,* 238–246.

Adams, J.W., & Hitch, G.J. (1997). Working memory and children's mental addition. *Journal of Experimental Child Psychology, 67,* 21–38.

Adams, J.W., & Hitch, G.J. (1998). Children's mental arithmetic and working memory. In C. Donlan (Ed.), *The development of mathematical skills* (pp. 153–173). Hove, East Sussex, United Kingdom: Psychology Press.

Advokat, C., Martino, L., Hill, B.D., & Gouvier, W. (2007). Continuous Performance Test (CPT) of college students with ADHD, psychiatric disorders, cognitive deficits, or no diagnosis. *Journal of Attention Disorders, 10,* 253–256.

Ahearn, E.M. (2008, July). State eligibility requirements for specific learning disabilities. *Project Forum at the National Association of State Directors of Special Education (NASDSE).* Retrieved November 3, 2008, from http://www.projectforum.org/docs/StateEligibilityRequirements forSpecificLearningDisabilities.pdf

Ahlberg, A., & Csocsán, E. (1999). How children who are blind experience numbers. *Journal of Visual Impairment and Blindness, 93,* 549–560.

Algozzine, B., O'Shea, D.J., Crews, W.B., & Stoddard, K. (1987). Analysis of mathematics competence of learning disabled adolescents. *The Journal of Special Education, 21,* 97–107.

Allison, T., McCarthy, G., Nobre, A., Puce, A., & Belger, A. (1994). Human extra-striate visual cortex and the perception of faces, words, numbers and colors. *Cerebral Cortex, 4,* 544–554.

Allsopp, D.H., Kyger, M.M., & Lovin, L.H. (2007) *Teaching mathematics meaningfully: Solutions for reaching struggling learners.* Baltimore: Paul H. Brookes Publishing Co.

American Psychiatric Association. (1980). *Diagnostic and statistical manual of mental disorders* (3rd ed.). Washington, DC: Author.

American Psychiatric Association. (2000). *Diagnostic and statistical manual of mental disorders* (4th ed., text rev.). Washington, DC: Author.

Andersson, M. (2003). Behavioural ecology: Coots count. *Nature, 422,* 483–485.

Ansari, D. (2007). Does the parietal cortex distinguish between "10," "ten," and ten dots? *Neuron, 53,* 165–167.

Ansari, D. (2008). Effects of development and enculturation on number representation in the brain. *Nature Reviews Neuroscience, 9,* 278–291.

Ansari, D., & Dhital, B. (2006). Age-related changes in the activation of the intraparietal sulcus during nonsymbolic magnitude processing: An event-related functional magnetic resonance imaging study. *Journal of Cognitive Neuroscience, 18,* 1820–1828.

Ansari, D., Garcia, N., Lucas, E., Hamon, K., & Dhital, B. (2005). Neural correlates of symbolic number processing in children and adults. *NeuroReport, 16,* 1769–1773.

Aram, D.M., Ekelman, B.L., & Nation, J.E. (1984). Preschoolers with language disorders ten years later. *Journal of Speech and Hearing Research, 27,* 232–244.

Arcavi, A. (1994, November). Symbol sense: Informal sense-making in formal mathematics. *For the Learning of Mathematics, 14*(3), 24–35.

Arcavi, A. (2003). The role of visual representations in the learning of mathematics. *Educational Studies in Mathematics, 52,* 215–241.

Ark, W.S. (2005). Comparing mental rotation and feature matching strategies in adults and children with behavioral and neuroimaging techniques. *Dissertation Abstracts International, 66*(09), 5112B.

Ashcraft, M.H., Donley, R.D., Halas, M.A., & Vakali, M. (1992). Working memory, automaticity, and problem difficulty. In J.I.D. Campbell (Ed.), *The nature and origins of mathematical skills* (pp. 301–329). Amsterdam: Elsevier.

Ashcraft, M.H., & Fierman, B.A. (1982). Mental addition in third, fourth, and sixth graders. *Journal of Experimental Child Psychology, 33,* 216–234.

Ashcraft, M.H., Kirk, E.P., & Hopko, D. (1998). On the cognitive consequences of mathematical anxiety. In C. Donlan (Ed.), *The development of mathematical skills* (pp. 175–196). Hove, East Sussex, United Kingdom: Psychology Press.

Ashcraft, M.H., Krause, J.A., & Hopko, D.R. (2007). Is math anxiety a mathematical learning disability? In D.B. Berch & M.M.M. Mazzocco (Eds.), *Why is math so hard for some children? The nature and origins of mathematical learning difficulties and disabilities* (pp. 329–348). Baltimore: Paul H. Brookes Publishing Co.

Ashcraft, M.H., & Stazyk, E.H. (1981). Mental addition: A test of three verification models. *Memory & Cognition, 9,* 185–196.

Bächtold, D., Baumuller, M., & Brugger, P. (1998). Stimulus-response compatibility in representational space. *Neuropsychologia, 36,* 731–735.

Baddeley, A. (2003). Working memory: Looking back and looking forward. *Nature Reviews Neuroscience, 4,* 829–839.

Badian, N.A. (1983). Arithmetic and non-verbal learning. In H.R. Myklebust (Ed.), *Progress in learning disabilities* (Vol. 5, pp. 235–264). New York: Grune and Stratton.

Badian, N.A. (2005). Does a visual-orthographic deficit contribute to reading disability? *Annals of Dyslexia, 55,* 28–52.

Bahrick, H.P., & Hall, L.K. (1991). Lifetime maintenance of high school mathematics content. *Journal of Experimental Psychology: General, 120,* 20–33.

Baillargeon, R. (1994). How do infants learn about the physical world? *Current Directions in Psychological Science, 3,* 133–140.

Baillargeon, R. (2002). The acquisition of physical knowledge in infancy: A summary of eight lessons. In U. Goswami (Ed.), *Blackwell handbook of childhood cognitive development* (pp. 47–83). Malden, MA: Blackwell.

Barbaresi, M.J., Katusic, S.K., Colligan, R.S., Weaver, A.L., & Jacobson, S.J. (2005). Math learning disorder: Incidence in a population-based birth cohort, 1976–1982, Rochester, Minn. *Ambulatory Pediatrics, 5,* 281–289.

Barkley, R.A. (1997). *ADHD and the nature of self-control.* New York: Guilford Press.

Barnes, M.A., Smith-Chant, B., & Landry, S.H. (2005). Number processing in neurodevelopmental disorders: Spina bifida myelomeningocele. In J.I.D. Campbell (Ed.), *Handbook of mathematical cognition* (pp. 299–313.). Hove, East Sussex, United Kingdom: Psychology Press.

Barnett, S.M., & Ceci, S.J. (2002). When and where do we apply what we learn?: A taxonomy for far transfer. *Psychological Bulletin, 128,* 612–637.

Baroody, A.J. (1987). *Children's mathematical thinking: A developmental framework for preschool, primary, and special education teachers.* New York: Teachers College Press.

Baroody, A.J. (2004). The role of psychological research in the development of early childhood mathematics standards. In D.H. Clements, J. Sarama (Eds.), & A.-M. DiBiase (Assoc. Ed.), *Engaging young children in mathematics: Standards for early childhood mathematics education* (pp. 149–172). Mahwah, NJ: Lawrence Erlbaum Associates.

Barriga, A.Q., Doran, J.W., Newell, S.B., Morrison, E.M., Barbetti, V., & Robbins, B.D. (2002). Relationships between problem behaviors and academic achievement in adolescents: The unique role of attention problems. *Journal of Emotional and Behavioral Disorders, 10,* 233–240.

Barrouillet, P., & Lepine, R. (2005). Working memory and children's use of retrieval to solve addition problems. *Journal of Experimental Child Psychology, 91,* 183–204.

Barth, H., Kanwisher, N., & Spelke, E. (2003). The construction of large number representations in adults. *Cognition, 86,* 201–221.

Barth, H., LaMont, K., Lipton, J., & Spelke, E.S. (2005). Abstract number and arithmetic in preschool children. *Proceedings of the National Academy of Sciences USA, 102,* 14116–14121.

Basso, A., Burgio, F., & Caporali, A. (2000). Acalculia, aphasia, and spatial disorders in left and right brain-damaged patients. *Cortex, 36,* 265–280.

Bassok, M. (2001). Semantic alignments in mathematical word problems. In D. Gentner, K.J. Holyoak, & B.N. Kokinov (Eds.), *The analogical mind: Perspectives from cognitive science.* Cambridge, MA: MIT Press.

Bassok, M. (2003). Analogical transfer in problem solving. In J.E. Davidson & R.J. Sternberg (Eds.), *The psychology of problem solving* (pp. 343–369). New York: Cambridge University Press.

Bassok, M., Chase, V.M., & Martin, S.A. (1998). Adding apples and oranges: Alignment of semantic and formal knowledge. *Cognitive Psychology, 35,* 99–134.

Bassok, M., Wu, L., & Olseth, K.L. (1995). Judging a book by its cover: Interpretive effects of content on problem solving transfer. *Memory & Cognition, 23,* 354–367.

Beck, A.T., Steer, R.A., & Brown, G.K. (1996). *Beck Depression Inventory–Second Edition (BDI-II).* San Antonio, TX: Psychological Corporation.

Beery, K.E., & Beery, N.A. (2004). *Beery-Buktenica Developmental Test of Visual-Motor Integration (VMI), Fifth Edition.* Minneapolis, MN: Pearson.

Beilock, S.L., Kulp, C.A., Holt, L.E., & Carr, T.H. (2004). More on the fragility of performance: Choking under pressure in mathematical problem solving. *Journal of Experimental Psychology: General, 133,* 584–600.

Beilstein, C.D., & Wilson, J.F. (2000). Landmarks in route learning by girls and boys. *Perceptual and Motor Skills, 91,* 877–882.

Benedetto-Nasko, E., & Tannock, R. (1999). Math computation, error patterns and stimulant effects in children with Attention Deficit Hyperactivity Disorder. *Journal of Attention Disorders, 3,* 121–134.

Bennett, C.M., & Baird, A.A. (2006). Anatomic changes in the emerging adult brain: A voxel-based morphometry study. *Human Brain Mapping, 27,* 766–777.

Bennett, G.K., Seashore, H.G., & Wesman, A.G. (1990). *Differential aptitude tests for personnel and career assessment.* San Antonio, TX: Harcourt.

Berch, D.B. (2005). Making sense of number sense: Implications for children with mathematical disabilities. *Journal of Learning Disabilities, 38,* 333–339.

Biederman, J., Petty, C., Fried, R., Fontanella, J., Doyle, A., Seidman, L.J., et al. (2006). Impact of psychometrically defined deficits of executive functioning in adults with Attention Deficit Hyperactivity Disorder. *American Journal of Psychiatry, 163,* 1730–1738.

Blackwell, L., Trzesniewski, K., & Dweck, C.S. (2007). Implicit theories of intelligence predict achievement across an adolescent transition: A longitudinal study of an intervention. *Child Development, 78,* 246–263.

Blajenkova, O., Kozhevnikov, M., & Motes, M.A. (2006). Object-spatial imagery: A new self-report imagery questionnaire. *Applied Cognitive Psychology, 20,* 239–263.

Bloom, B.S., Hastings, J.T., & Madaus, G.F. (1971) *Handbook on formative and summative evaluation of student learning.* New York: McGraw-Hill.

Blöte, A.W., Van der Burg, E., & Klein, A.S. (2001). Students' flexibility in solving two-digit addition and subtraction problems: Instructional effects. *Journal of Educational Psychology, 93,* 627–638.

Boaler, J. (2000). Exploring situated insights into research and learning. *Journal for Research in Mathematics Education, 31,* 113–119.

Booksh, R.L. (2005). *Ability of college students to simulate ADHD on objective measures of attention.* Unpublished doctoral dissertation, Louisiana State University, Baton Rouge.

Boone, H.C. (1986). Relationship of left-right reversals to academic achievement. *Perceptual and Motor Skills, 62,* 27–33.

Booth, J.L., & Siegler, R.S. (2006). Developmental and individual differences in pure numerical estimation. *Developmental Psychology, 41,* 189–201.

Booth, J.L., & Siegler, R.S. (2008). Numerical magnitude representations influence children's arithmetic learning. *Child Development, 79,* 1016–1031.

Bormuth, J.R. (1966). Readability: A new approach. *Reading Research Quarterly, 1,* 79–132.

Boroditsky, L. (2000). Metaphoric structuring: Understanding time through spatial metaphors. *Cognition, 75,* 1–28.

Brainerd, C.J., & Gordon, L.L. (1994). Development of verbatim and gist memory for numbers. *Developmental Psychology, 30,* 163–177.

Brannon, E.M., & Van de Walle, C.T. (2001). The development of ordinal numerical competence in young children. *Cognitive Psychology, 43,* 53–81.

Bryant, D.P., Bryant, B.R., & Hammill, D.D. (2000). Characteristic behaviors of students with LD who have teacher-identified math weaknesses. *Journal of Learning Disabilities, 33,* 168–177, 199.

Bull, R., Johnston, R.S., & Roy, J.A. (1999). Exploring the roles of the visual-spatial sketch pad and central executive in children's arithmetic skills: Views from cognition and developmental neuropsychology. *Developmental Neuropsychology, 15,* 421–442.

Bull, R., Marschark, M., & Blatto-Vallee, G. (2005). SNARC hunting: Examining number representation in deaf students. *Learning and Individual Differences, 15,* 223–236.

Bull, R., & Scerif, G. (2001). Executive functioning as a predictor of children's mathematics ability: Inhibition, switching, and working memory. *Developmental Neuropsychology, 19,* 273–293.

Burnett, S.A., Lane, D.L., & Dratt, L.M. (1979). Spatial visualization and sex differences in quantitative ability. *Intelligence, 3,* 345–354.

Butterworth, B. (2003). *Dyscalculia Screener.* London: NFER Nelson.

Butterworth, B., Cappelletti, M., & Kopelman, M. (2001). Category specificity in reading and writing: The case of number words. *Nature Neuroscience, 4,* 784–786.

Butterworth, B., Marchesini, N., & Girelli, L. (2003). Multiplication facts: Passive storage or dynamic reorganization? In A.J. Baroody & A. Dowker (Eds.), *The development of arithmetic concepts and skills* (pp. 189–202). Mahwah, NJ: Lawrence Erlbaum Associates.

Butterworth, B., Zorzi, M., Girelli, L., & Jonckheere, A.R. (2001). Storage and retrieval of addition facts: The role of number comparison. *The Quarterly Journal of Experimental Psychology. A: Human Experimental Psychology, 54,* 1005–1029.

Caldwell, J.H., & Golin, G.A. (1987). Variables affecting word problem difficulty in secondary school mathematics. *Journal for Research in Mathematics Education, 18,* 187–196.

Campbell, J.I.D., & Xue, Q. (2001). Cognitive arithmetic across cultures. *Journal of Experimental Psychology: General, 130,* 299–315.

Campbell, S. (2005). Three philosophical perspectives on logic and psychology: Implications for mathematics education. *Philosophy of Mathematics Education Journal 14.* Retrieved on May 19, 2007, from http://www.people.ex.ac.uk/PErnest/pome14/campbell.htm

Canobi, K.H., Reeve, R.A., & Pattison, P.E. (1998). The role of conceptual understanding in children's addition problem solving. *Developmental Psychology, 34,* 882–891.

Cantlon, J.F., Brannon, E.M., Carter, E.J., Pelphrey, K.A. (2006, May). Functional imaging of numerical processing in adults and 4-y-old children. *Public Library of Science Biology, 4*(5), e125.

Cappelletti, M., Butterworth, B., & Kopelman, M. (2001). Spared numerical abilities in a case of semantic dementia. *Neuropsychologia, 39,* 1224–1239.

Carpenter, P.A., & Just, M.A. (1986). Spatial ability: An information processing approach to psychometrics. In R.J. Sternberg (Ed.), *Advances in the psychology of human intelligence* (Vol. 3, pp. 221–253). Mahwah, NJ: Lawrence Erlbaum Associates.

Carpenter, P.A., Just, M.A., & Reichle, E.D. (2000). Working memory and executive function: Evidence from neuroimaging. *Current Opinion in Neurobiology, 10,* 195–199.

Carpenter, T.P., Corbitt, M.K., Kepner, H.S., Jr., Lindquist, M.M., & Reys, R.E. (1980). Solving verbal problems: Results and implications from national assessment. *Arithmetic Teacher, 28*(1), 8–12.

Carpenter, T.P., Corbitt, M.K., Kepner, H.S., Lindquist, M.M., & Reys, R.E. (1981). *Results from the second mathematics assessment of the National Assessment of Educational Progress*. Washington, DC: National Council of Teachers of Mathematics.

Carpenter, T.P., Fennema, E., Franke, M.L., Levi, L., & Empson, S.B. (1999). *Children's mathematics: Cognitively guided instruction*. Portsmouth, NH: Heinemann.

Carpenter, T.P., Hiebert, J., & Moser, J.M. (1983). The effect of instruction on children's solution of addition and subtraction word problems. *Educational Studies in Mathematics, 14*, 55–72.

Carroll, L. (1872/1960a). Alice's adventures in wonderland. In M. Gardner, *The annotated Alice*. New York: The World Publishing Company.

Carroll, L. (1872/1960b). Through the looking-glass. In M. Gardner, *The annotated Alice*. New York: The World Publishing Company.

Carroll, W.M. (1994). Using worked examples as an instructional support in the algebra classroom. *Journal of Educational Psychology, 86*, 360–367.

Case, R., & Griffin, S. (1990). Child cognitive development: The role of central conceptual structures in the development of scientific and social thought. In C.-A. Hauert (Ed.), *Developmental psychology: Cognitive, perceptual-motor and psychological perspectives* (pp. 193–230). Amsterdam: Elsevier.

Casey, B.J., Giedd, J.N., & Thomas, K.M. (2000). Structural and functional brain development and its relation to cognitive development. *Biological Psychology, 54*, 241–257.

Casey, M.B., & Brabeck, M.M. (1989). Exceptions to the male advantage on a spatial task: Family handedness and college major as a factor identifying women who excel. *Neuropsychologia, 27*, 689–696.

Casey, M.B., Nuttall, R.L., & Pezaris, E. (1997). Mediators of gender differences in mathematics college entrance test scores: A comparison of spatial skills with internalized beliefs and anxieties. *Developmental Psychology, 33*, 669–680.

Casey, M.B., Nuttall, R.L., & Pezaris, E. (2001). Spatial-mechanical reasoning skills versus mathematics self-confidence as mediators of gender differences on mathematics subtests using cross-national gender-based items. *Journal for Research in Mathematics Education, 32*, 28–57.

Cattell, R.B. (1971). *Abilities: Their structure, growth and action*. Boston: Houghton Mifflin.

Chan, J. (1981). A crossroads in language instruction. *Journal of Reading, 24*, 411–415.

Chard, D., Clarke, B., Baker, S., Otterstedt, J., Braun, D., & Katz, R. (2005). Using measures of number sense to screen for difficulties in mathematics: Preliminary findings. *Assessment for Effective Intervention, 30*, 3–14.

Chase, W.G., & Simon, H.A. (1973). Perception in chess. *Cognitive Psychology, 4*, 55–81.

Cheng, K., & Newcombe, N.S. (2005). Is there a geometric module for spatial orientation? Squaring theory and evidence. *Psychonomic Bulletin & Review, 12*, 1–23.

Cheng, L., & Huntley-Fenner, G. (2000, March). *How do native Chinese speakers treat count-mass in English?* Paper presented at the Language, Culture, and Cognition Meeting, Leiden University, Netherlands.

Chiappe, P. (2005). How reading research can inform mathematics difficulties: The search for the core deficit. *Journal of Learning Disabilities, 38*, 313–317.

Chinn, S., & Ashcroft, R. (2007). *Mathematics for dyslexics, including dyscalculia*. West Sussex, England: Wiley.

Chochon, F., Cohen, L., van de Moortele, P.-F., & Dehaene, S. (1999). Differential contributions of the left and right inferior parietal lobules to number processing. *Journal of Cognitive Neuroscience, 11*, 617–630.

Christensen, C.A., & Gerber, M.M. (1990). Effectiveness of computerized drill and practice games in teaching basic mathematics facts. *Exceptionality, 1*, 149–165.

Church, R.B. (1999). Using gesture and speech to capture transitions in learning. *Cognitive Development, 14*, 313–342.

Cipolotti, L., Warrington, E.K., & Butterworth, B. (1995). Selective impairment in manipulating Arabic numerals. *Cortex, 31*, 73–86.

Clark, D.B., & Uhry, J.K. (1995). *Dyslexia: Theory and practice of remedial instruction.* Baltimore: York Press.

Clarren, S.B., Martin, D.C., & Townes, B.D. (1993). Academic achievement over a decade: A neuropsychological prediction study. *Developmental Neuropsychology, 9,* 161–176.

Clement, J., Lochhead, J., & Monk, G.S. (1981). Translation difficulties in learning mathematics. *American Mathematical Monthly, 88,* 286–290.

Clements, D.H. (2004). Geometric and spatial thinking in early childhood education. In D.H. Clements, J. Sarama (Eds.), & A.-M. DiBiase (Assoc. Ed.), *Engaging young children in mathematics: Standards for early childhood mathematics education* (pp. 267–297). Mahwah, NJ: Lawrence Erlbaum Associates.

Cocude, M., Mellet, E., & Denis, M. (1999). Visual and mental exploration of visuo-spatial configurations: Behavioral and neuroimaging approaches. *Psychological Research, 62,* 93–106.

Cohen, L., & Dehaene, S. (1996). Cerebral networks for number processing: Evidence from a case of posterior callosal lesion. *NeuroCase, 2,* 155–174.

Cohen, L., & Dehaene, S. (2004). Specialization within the ventral stream: The case for the visual word form area. *NeuroImage, 22,* 466–476.

Cohen, L., Dehaene, S., Chochon, F., Lehéricy, S., & Naccache, L. (2000). Language and calculation within the parietal lobe: A combined cognitive, anatomical and fMRI study. *Neuropsychologia, 38,* 1426–1440.

Cohen, L., Verstichel, P., & Dehaene, S. (1997). Neologistic jargon sparing numbers: A category-specific phonological impairment. *Cognitive Neuropsychology, 14,* 1029–1061.

College Board (2007). *2007 College-bound seniors: Total group profile report.* Retrieved on January 9, 2008, from http://professionals.collegeboard.com/data-reports-research/SAT/cb-seniors-2007

Conners, C.K. (2008). *Conners 3rd Edition (Conners 3).* North Tonawanda, NY: Multi-Health Systems.

Connolly, A.J. (2007). *KeyMath 3.* Minneapolis, MN: Pearson.

Corballis, M.C. (1997). Mental rotation and the right hemisphere. *Brain and Language, 57,* 100–121.

Cowan, R., Donlan, C., Newton, E.J., & Lloyd, D. (2005). Number skills and knowledge in children with specific language impairment. *Journal of Educational Psychology, 97,* 732–744.

Culham, J.C., & Kanwisher, N.G. (2001). Neuroimaging of cognitive functions in human parietal cortex. *Current Opinion in Neurobiology, 11,* 157–163.

Cummins, D.D., Kintsch, W., Reusser, K., & Weimer, R. (1988). The role of understanding in solving word problems. *Cognitive Psychology, 20,* 405–438.

Dark, V.J., & Benbow, C.P. (1991). Differential enhancement of working memory with mathematical versus verbal precocity. *Journal of Educational Psychology, 83,* 48–60.

Davis, H., & Pérusse, R. (1988). Numerical competence in animals: Definitional issues, current evidence, and a new research agenda. *Behavioral and Brain Sciences, 11,* 561–615.

Davis, S.H. (2007). Bridging the gap between research and practice: What's good, what's bad, and how can one be sure? *Phi Delta Kappan, 88,* 569–578.

Dawe, L. (1983). Bilingualism and mathematical reasoning in English as a second language. *Educational Studies in Mathematics, 14,* 325–353.

Deater-Deckard, K., Petrill, S.A., Thompson, L.A., & DeThorne, L. (2005). A cross-sectional behavioural genetic analysis of task persistence in the transition to middle childhood. *Developmental Science, 8,* F21–F26.

DeCorte, E., Verschaffel, L., & DeWin, L. (1985). Influence of rewording verbal problems on children's problem representations and solutions. *Journal of Educational Psychology, 77,* 460–470.

Dehaene, S. (1992). Varieties of numerical abilities. *Cognition, 44,* 1–42.

Dehaene, S. (1997). *The number sense: How the mind creates mathematics.* New York: Oxford University Press.

Dehaene, S. (1999). Fitting two languages into one brain [Editorial]. *Brain, 122,* 2207–2208.

Dehaene, S. (2000). Cerebral bases of number processing and calculation. In M.S. Gazzaniga (Ed.), *The new cognitive neurosciences* (2nd ed., pp. 987–998). Cambridge, MA: MIT Press.

Dehaene, S. (2005). Evolution of human cortical circuits for reading and arithmetic: The "neuronal recycling" hypothesis. In S. Dehaene, J.R. Duhamel, M. Hauser, & G. Rizzolatti (Eds.), *From monkey brain to human brain* (pp. 133–158). Cambridge, MA: MIT Press.

Dehaene, S., Bossini, S., & Giraux, P. (1993). The mental representation of parity and number magnitude. *Journal of Experimental Psychology: General, 122,* 371–396.

Dehaene, S., & Cohen, L. (1991). Two mental calculation systems: A case study of severe acalculia with preserved approximation. *Neuropsychologia, 29,* 1045–1074.

Dehaene, S., Cohen, L., & Changeux, J.P. (1998). Neuronal network models of acalculia and prefrontal deficits. In R.W. Parks, D.S. Levine, & D.L. Long (Eds.), *Fundamentals of neural network modeling: Neuropsychology and cognitive neuroscience* (pp. 233–256). Cambridge, MA: MIT Press.

Dehaene, S., Cohen, L., Sigman, M., & Vinckier, F. (2005). The neural code for written words: A proposal. *Trends in Cognitive Sciences, 9,* 335–341.

Dehaene, S., Izard, V., Pica, P., & Spelke, E. (2006, January 20). Core knowledge of geometry in an Amazonian indigene group. *Science, 311,* 381–384.

Dehaene, S., Molko, N., Cohen, L., & Wilson, A.J. (2004). Arithmetic and the brain. *Current Opinion in Neurobiology, 14,* 218–224.

Dehaene, S., Piazza, M., Pinel, P., & Cohen, L. (2003). Three parietal circuits for number processing. *Cognitive Neuropsychology, 20,* 487–506.

Dehaene, S., Spelke, E., Pinel, P., Stanescu, R., & Tsivkin, S. (1999, May 7). Sources of mathematical thinking: Behavioral and brain-imaging evidence. *Science, 284,* 970–974.

Delazer, M., Girelli, L., Semenza, C., & Denes, G. (1999). Numerical skills and aphasia. *Journal of the International Neuropsychological Society, 5,* 213–221.

Delazer, M., Ischebeck, A., Domahs, F., Zamarian, L., Koppelstaetter, F., Siedentopf, C.M., et al. (2005). Learning by strategies and learning by drill—Evidence from an fMRI study. *NeuroImage, 25,* 838–849.

Deloche, G., Seron, X., Larroque, C., Magnien, C., Metz-Lotz, M.N., Noel, M.N., et al. (1994). Calculation and number processing assessment battery: The role of demographic factors. *Journal of Clinical and Experimental Neuropsychology, 16,* 195–208.

Delson, R.M. (2006, June 2). In their report "Core knowledge of geometry in an Amazonian indigene group" [Letter to the editor]. *Science, 312,* 1309–1320.

Dembo, Y., Levin, I., & Siegler, R.S. (1997). A comparison of the geometric reasoning of students attending Israeli ultraorthodox and mainstream schools. *Developmental Psychology, 33,* 92–103.

Desoete, A., & Grégoire, J. (2006). Numerical competence in young children and in children with mathematics learning disabilities. *Learning and Individual Differences, 16,* 351–367.

Desoete, A., Roeyers, H., & DeClercq, A. (2004). Children with mathematics learning disabilities in Belgium. *Journal of Learning Disabilities, 37,* 50–61.

Detterman, D.K. (1993). The case for the prosecution: Transfer as an epiphenomenon. In D.K. Detterman & R.J. Sternberg (Eds.), *Transfer on trial: Intelligence, cognition, and instruction* (pp. 1–24). Norwood, NJ: Ablex.

Devenport, J.A., Patterson, M.R., & Devenport, L.D. (2005). Averaging and foraging decisions in horses *(Equus calabus). Journal of Comparative Psychology, 119,* 352–358.

Doerr, H.M., & Tinto, P.P. (2000). Paradigms for teacher-centered, classroom-based research. In A.E. Kelly & R.A. Lesh (Eds.), *Handbook of research design in mathematics and science education* (pp. 403–427). Mahwah, NJ: Lawrence Erlbaum Associates.

Donlan, C. (1998). Number without language? Studies of children with specific language impairments. In C. Donlan (Ed.), *The development of mathematical skills* (pp. 255–274). Hove, East Sussex, United Kingdom: Psychology Press.

Donlan, C., & Gourlay, S. (1999). The importance of non-verbal skills in the acquisition of place-value knowledge: Evidence from normally developing and language-impaired children. *British Journal of Developmental Psychology, 17,* 1–19.

Douglas, V.I., Barr, R.G., O'Neill, M.E., & Britton, B.G. (1986). Short term effects of methylphenidate on the cognitive learning and academic performance of children with attention deficit disorder in the laboratory and the classroom. *Journal of Child Psychology and Psychiatry, 27,* 191–212.

Dowker, A.D. (1992). Computational estimation strategies of professional mathematicians. *Journal for Research in Mathematics Education, 23,* 45–55.

Dowker, A. (1997). Young children's addition estimates. *Mathematical Cognition, 3,* 141–154.

Dowker, A. (2005). Early identification and intervention for students with mathematics difficulties. *Journal of Learning Disabilities, 38,* 324–332.

Dowker, A. (2008). Individual differences in numerical abilities in preschoolers. *Developmental Science, 11,* 650–654.

Doyle, A.E., Wilens, T.E., Kwon, A., Seidman, L.J., Faraone, S.V., Fried, R., et al. (2005). Neuropsychological functioning in youth with bipolar disorder. *Biological Psychiatry, 58,* 540–548.

Duffau, H., Denvil, D., Lopes, M., Gasparini, F., Cohen, L., Capelle, L., et al. (2002). Intraoperative mapping of the cortical areas involved in multiplication and subtraction: An electrostimulation study in a patient with a left parietal glioma. *Journal of Neurology, Neurosurgery, and Psychiatry, 73,* 733–738.

Duffy, G.G., & Kear, K. (2007). Compliance or adaptation: What is the real message about research-based practices? *Phi Delta Kappan, 88,* 579–581.

Dunbar, K. (2001). The analogical paradox: Why analogy is so easy in naturalistic settings, yet so difficult in the psychological laboratory. In D. Gentner, K. Holyoak, & B. Kokinov (Eds.), *Analogy: Perspectives from cognitive science* (pp. 313–334). Cambridge, MA: MIT Press.

Duncan, G.J., Dowsett, C.J., Claessens, A., Magnuson, K., Huston, A.C., Klebanov, P., et al. (2007). School readiness and later achievement. *Developmental Psychology, 43,* 1428–1446.

Durand, M., Hulme, C., Larkin, R., & Snowling, M. (2005). The cognitive foundations of reading and arithmetic skills in 7- to 10-year-olds. *Journal of Experimental Child Psychology, 91,* 113–136.

Durkin, K., Shire, B., Riem, R., Crowther, R.D., & Rutter, D.R. (1986). The social and linguistic context of early number word use. *British Journal of Developmental Psychology, 4,* 269–288.

Dykens, E., Leckman, J., Riddle, M., Hardin, M., Schwartz, S., & Cohen, D. (1990). Intellectual, academic, and adaptive functioning of Tourette syndrome children with and without attention deficit disorder. *Journal of Abnormal Child Psychology, 18,* 607–615.

Ebersbach, M., Luwel, K., Frick, A., Onghena, P., & Verschaffel, L. (2008). The relationship between the shape of the mental number line and familiarity with numbers in 5- to 9-year-old children: Evidence for a segmented linear model. *Journal of Experimental Child Psychology, 99,* 1–17.

Eger, E., Sterzer, P., Russ, M.O., Giraud, A.L., & Kleinschmidt, A. (2003). A supramodal number representation in human intraparietal cortex. *Neuron, 37,* 719–725.

Ekenstam, A., & Greger, K. (1983). Some aspects of children's ability to solve mathematical problems. *Educational Studies in Mathematics, 14,* 369–384.

Ellis, N.C., & Miles, T.R. (1977). Dyslexia as a limitation in the ability to process information. *Bulletin of the Orton Society, 27,* 72–81.

Espy, K.A., McDiarmid, M.M., Cwik, M.F., Stalets, M.M., Hamby, A., & Senn, T.E. (2004). The contribution of executive functions to emergent mathematic skills in preschool children. *Developmental Neuropsychology, 26,* 465–486.

Eves, H. (2003). *Mathematical circles* (Vol. 1). Washington, DC: The Mathematical Association of America.

Faraone, S.V., Biederman, J., Lehman, B.K., Spencer, T., Norman, D., Seidman, L.J., et al. (1993). Intellectual performance and school failure in children with attention deficit hyperactivity disorder and in their siblings. *Journal of Abnormal Psychology, 102,* 616–623.

Fazio, B. (1996). Mathematical abilities of children with specific language impairment: A follow-up study. *Journal of Speech and Hearing Research, 39,* 839–849.

Feldman, A., & Minstrell, J. (2000). Action research as a research methodology for the study of the teaching and learning of science. In A.E. Kelly & R.A. Lesh (Eds.), *Handbook of research design in mathematics and science education* (pp. 429–455). Mahwah, NJ: Lawrence Erlbaum Associates.

Ferrini-Mundy, J. (1987). Spatial training for calculus students: Sex differences in achievement and in visualization ability. *Journal for Research in Mathematics Education, 18,* 126–140.

Ferrini-Mundy, J., & Lauten, D. (1993). Teaching and learning calculus. In P.S. Wilson (Ed.), *Research ideas for the classroom: High school mathematics* (pp. 155–176). New York: Macmillan.

Ferro, J.M., & Botelho, S. (1980). Alexia for arithmetical signs. A cause of disturbed calculation. *Cortex, 16,* 175–180.

Fias, W., & Fischer, M.H. (2005). Spatial representation of numbers. In J.I.D. Campbell (Ed.), *Handbook of mathematical cognition* (pp. 43–54). New York: Psychology Press.

Fias, W., Lammertyn, J., Reynvoet, B., Dupont, P., & Orban, G.A. (2003). Parietal representation of symbolic and nonsymbolic magnitude. *Journal of Cognitive Neuroscience, 15,* 47–56.

Fischer, F.E., & Beckey, R.D. (1990). Beginning kindergartener's perception of number. *Perceptual and Motor Skills, 70,* 419–425.

Fischer, M.H. (2003). Cognitive representation of negative numbers. *Psychological Science, 14,* 278–282.

Fisher, C., Klingler, S.L., & Song, H.-J. (2006). What does syntax say about space? 2-year-olds use sentence structure to learn new prepositions. *Cognition, 101,* B19–B29.

Fisher, D., & Frey, N. (2007). *Checking for understanding: Formative assessment techniques for your classroom.* Alexandria, VA: Association for Curriculum Supervision and Development.

Flanagan, D.P., Ortiz, S.O., & Alfonso, V.C. (2007). *Essentials of cross-battery assessment* (2nd ed.). Hoboken, NJ: Wiley.

Fleischner, J.E. (1994). Diagnosis and assessment of mathematics learning disabilities. In G.R. Lyon (Ed.), *Frames of reference for the assessment of learning disabilities: New views on measurement issues* (pp. 441–458). Baltimore: Paul H. Brookes Publishing Co.

Floyd, R.G., Evans, J.J., & McGrew, K.S. (2003). Relations between measures of Cattell-Horn-Carroll (CHC) cognitive abilities and mathematics achievement across the school-age years. *Psychology in the Schools, 40,* 155–171.

Francis, D.J., Fletcher, J.M., Stuebing, K.K., Lyon, G.R., Shaywitz, B.A., & Shaywitz, S.E. (2005). Psychometric approaches to the identification of LD: IQ and achievement scores are not sufficient. *Journal of Learning Disabilities, 38,* 98–108.

Franco, L., & Sperry, R.W. (1977). Hemisphere lateralization for cognitive processing of geometry. *Neuropsychologia, 15,* 107–113.

Frank, M.C., Everett, D.L., Fedorenko, E., & Gibson, E. (2008). Number as a cognitive technology: Evidence from Pirahã language and cognition. *Cognition, 108,* 819–824.

Frauenheim, J.G. (1978). Dyslexic children as adults. *Journal of Learning Disabilities, 11,* 476–483.

French-Mestre, C., & Vaid, J. (1993). Activation of number facts in bilinguals. *Memory & Cognition, 21,* 809–818.

Friedman, L. (1995). The space factor in mathematics: Gender differences. *Review of Educational Research, 65,* 22–50.

Fuchs, L.S., Compton, D.L., Fuchs, D., Paulsen, K., Bryant, J.D., & Hamlett, C.L. (2005). The prevention, identification, and cognitive determinants of math difficulty. *Journal of Educational Psychology, 97,* 493–513.

Fuchs, L.S., & Fuchs, D. (2002). Mathematical problem-solving profiles of students with mathematical disabilities with and without comorbid reading disabilities. *Journal of Learning Disabilities, 35,* 563–573.

Fuchs, L.S., Fuchs, D., Compton, D.L., Powell, S.R., Seetahler, P.M., Capizzi, A.M., et al. (2006). The cognitive correlates of third-grade skill in arithmetic, algorithmic computation, and arithmetic word problems. *Journal of Educational Psychology, 98,* 29–43.

Fuchs, L.S., Fuchs, D., Finelli, R., Courey, S.J., Hamlett, C.L., Sones, E.M., et al. (2006). Teaching third graders about real-life mathematical problem solving: A randomized controlled study. *The Elementary School Journal, 106,* 293–311.

Fueyo, V., & Bushell, D. (1998). Using number-line procedures and peer tutoring to improve the mathematics computation of low-performing first graders. *Journal of Applied Behavior Analysis, 31,* 417–430.

Fürst, A.J., & Hitch, G.J. (2000). Separate roles for executive and phonological components of working memory in mental arithmetic. *Memory & Cognition, 28,* 774–782.

Fuson, K.C. (1988). *Children's counting and concepts of number.* New York: Springer-Verlag.

Fuson, K.C., Smith, S.T., & LoCicero, A.M. (1997). Supporting Latino first graders' ten-structured thinking in urban classrooms. *Journal for Research in Mathematics Education, 28,* 738–766.

Fuson, K.C., & Willis, G.B. (1989). Second graders' use of schematic drawings in solving addition and subtraction word problems. *Journal of Educational Psychology, 81,* 514–520.

Galea, L.A., & Kimura, D. (1993). Sex differences in route learning. *Personality and Individual Differences, 14,* 53–65.

Gallagher, A.M. (1992). *Sex differences in problem-solving used by high-scoring examinees on the SAT-M* (College Board Report No. 92-2, Educational Testing Service Research Report No. RR-92-33). New York: College Board Publications.

Gallagher, A.M., DeLisi, R., Holst, P.C., McGillicuddy-DeLisi, A.V., Morley, M., & Cahalan, C. (2000). Gender differences in advanced mathematical problem solving. *Journal of Experimental Child Psychology, 75,* 165–190.

Gallagher, A.M., Levin, J., & Cahalan, C. (2002, September). *Cognitive patterns of gender differences on mathematics admissions tests* (Graduate Record Examinations Board Report No. 96-17P, Educational Testing Service Research Report No. RR-02-19). Retrieved on April 6, 2006, from www.ets.org/Media/Research/pdf/RR-02-19-Gallagher.pdf

Gazzaniga, M.S. (1995). Principles of human brain organization derived from split-brain studies. *Neuron, 14,* 217–228.

Geary, D.C. (1993). Mathematical disabilities: Cognitive, neuropsychological, and genetic components. *Psychological Bulletin, 114,* 345–362.

Geary, D.C. (1996). The problem size effect in mental addition: Developmental and cross-national trends. *Mathematical Cognition, 2,* 63–93.

Geary, D.C. (2003). Learning disabilities in arithmetic: Problem-solving differences and cognitive deficits. In H.L. Swanson, K.R. Harris, & S. Graham (Eds.), *Handbook of learning disabilities* (pp. 199–212). New York: Guilford Press.

Geary, D.C., Berch, D.B., Boykin, A.W., Embretson, S., Reyna, V., Siegler, R., et al. (2008). *Report of the Task Group on Learning Processes. National Mathematics Advisory Panel.* Retrieved on March 15, 2008, from http://www.ed.gov/about/bdscomm/list/mathpanel/index.html

Geary, D.C., Bow-Thomas, C.C., & Yao, Y. (1992). Counting knowledge and skill in cognitive addition: A comparison of normal and mathematically disabled children. *Journal of Experimental Child Psychology, 54,* 372–391.

Geary, D.C., & DeSoto, M.C. (2001). Sex differences in spatial abilities among adults from the United States and China. *Evolution and Cognition, 7,* 172–177.

Geary, D.C., Fan, L., & Bow-Thomas, C.C. (1992). Numerical cognition: Locus of ability differences comparing children from China and the United States. *Psychological Science, 3,* 180–185.

Geary, D.C., Hoard, M.K., Byrd-Craven, J., DeSoto, M.C. (2004). Strategy choices in simple and complex addition: Contributions of working memory and counting knowledge for children with mathematical disability. *Journal of Experimental Child Psychology, 88,* 121–151.

Geary, D.C., Hoard, M.K., Byrd-Craven, J., Nugent, L., & Numtee, C. (2007). Cognitive mechanisms underlying achievement deficits in children with mathematical learning disability. *Child Development, 78,* 1343–1359.

Geary, D.C., Hoard, M.K., Nugent, L., Byrd-Craven, J. (2008). Development of number line representations in children with mathematical learning disability. *Developmental Neuropsychology, 33,* 277–299.

Gelman, R., & Butterworth, B. (2005). Number and language: How are they related? *Trends in Cognitive Sciences, 9,* 6–10.

Gelman, R., Durgin, F., & Kaufman, L. (1995). Distinguishing between animates and inanimates: Not by motion alone. In D. Sperber, D. Premack, & A.J. Premack (Eds.), *Causal cognition: A multidisciplinary debate* (pp. 150–184). Oxford, United Kingdom: Clarendon.

Gelman, R., & Gallistel, C.R. (1978/1986). *The child's understanding of number* (rev. ed.). Cambridge, MA: Harvard University Press.

Gersten, R., Ferrini-Mundy, J., Benbow, C., Clements, D.H., Loveless, T., Williams, V., et al. (2008). *Report of the Task Group on Instructional Practices. National Mathematics Advisory Panel.*

Retrieved on March 15, 2008, from http://www.ed.gov/about/bdscomm/list/mathpanel/report/instructional-practices.pdf

Gersten, R., Jordan, N.C., & Flojo, J.R. (2005). Interventions for students with mathematics difficulties. *Journal of Learning Disabilities, 38,* 293–304.

Gevers, W., & Lammertyn, J. (2005). The hunt for SNARC. *Psychology Science, 47,* 10–21.

Gevers, W., Reynvoet, B., & Fias, W. (2003). The mental representation of ordinal sequences is spatially organised. *Cognition, 87,* B87–B95.

Ginsburg, H.P., & Baroody, A.J. (2003). *Test of Early Mathematics Ability–Third Edition (TEMA-3).* Austin, TX: PRO-ED.

Ginsburg, H.P., Jacobs, S.F., & Lopez, L. (1993). Assessing mathematical thinking and learning potential. In R.B. Davis & C.A. Maher (Eds.), *Schools, mathematics, and the world of reality* (pp. 237–262). Boston: Allyn & Bacon.

Ginsburg, H.P., Posner, J.K., & Russell, R.L. (1981). The development of mental addition as a function of schooling and culture. *Journal of Cross-Cultural Psychology, 12,* 163–178.

Gioia, G.A., Isquith, P.K., Guy, S.C., & Kenworthy, L. (2000). *Behavior Rating Inventory of Executive Function (BRIEF)—Parent and Teacher Forms.* Lutz, FL: Psychological Assessment Resources.

Girelli, L., Lucangeli, D., & Butterworth, B. (2000). The development of automaticity in accessing number magnitude. *Journal of Experimental Child Psychology, 76,* 104–122.

Göbel, S.M., Calabria, M., Farné, A., & Rossetti, Y. (2006). Partial rTMS distorts the mental number line: Simulating 'spatial' neglect in healthy subjects. *Neuropsychologia, 44,* 860–868.

Goel, V., & Dolan, R.J. (2004). Differential involvement of left prefrontal cortex in inductive and deductive reasoning. *Cognition, 93,* B109–B121.

Goldin-Meadow, S., Nusbaum, H., Kelly, S.D., & Wagner, S. (2001). Explaining math: Gesturing lightens the load. *Psychological Science, 12,* 516–522.

Gonzales, P., Guzmán, J.C., Partelow, L., Pahlke, E., Jocelyn, L., Kastberg, D., et al. (2004). *Highlights from the Trends in International Mathematics and Science Study (TIMSS) 2003* (NCES 2005-005). U.S. Department of Education, National Center for Education Statistics. Washington, DC: U.S. Government Printing Office.

Gonzalez, J.E.J., & Espinel, A.I.G. (1999). Is IQ-achievement discrepancy relevant in the definition of arithmetic learning disabilities? *Learning Disability Quarterly, 22,* 291–301.

Goswami, U. (2001). Analogical reasoning in children. In D. Gentner, K.J. Holyoak, & B.N. Kokinov (Eds.), *The analogical mind: Perspectives from cognitive science* (pp. 437–470). Cambridge, MA: MIT Press.

Goswami, U., & Brown, A.L. (1990). Higher-order structure and relational reasoning: Contrasting analogical and thematic relations. *Cognition, 36,* 207–226.

Gottfredson, L.S. (1997). Mainstream science on intelligence: An editorial with 52 signatories, history, and bibliography. *Intelligence, 24,* 13–23.

Grabner, R.H., Ansari, D., Reishofer, G., Stern, E., Ebner, F., & Neuper, C. (2007). Individual differences in mathematical competence predict parietal brain activation during mental calculation. *NeuroImage, 38,* 346–356.

Grafman, J. (2002). The human prefrontal cortex has evolved to represent components of structured event complexes. In J. Grafman (Ed.), *The frontal lobes* (pp. 157–175). Amsterdam: Elsevier.

Graham, A.T., & Thomas, M.W. (2000). Building a versatile understanding of algebraic variables with a graphic calculator. *Educational Studies in Mathematics, 41,* 265–282.

Graham, S., & Golan, S. (1991). Motivational influences on cognition: Task involvement, ego involvement, and depth of information processing. *Journal of Educational Psychology, 83,* 187–194.

Gravemeijer, K. (1994). Educational development and developmental research in mathematics education. *Journal for Research in Mathematics Education, 25,* 443–471.

Gray, J.R., Chabris, C., & Braver, T. (2003). Neural mechanisms of general fluid intelligence. *Nature Neuroscience, 6,* 316–322.

Graziano, A.B., Peterson, M., & Shaw, G.L. (1999). Enhancing learning of proportional math through music training and spatial-temporal training. *Neurological Research, 21,* 139–152.

Gregory, J.W., & Osborne, A.R. (1975). Logical reasoning ability and teacher verbal behavior within the mathematics classroom. *Journal for Research in Mathematics Education, 6*, 26–36.

Griffin, S. (2004). Number Worlds: A research-based mathematics program for young children. In D.H. Clements, J. Sarama (Eds.), & A.-M. DiBiase (Assoc. Ed.), *Engaging young children in mathematics: Standards for early childhood mathematics education* (pp. 325–340). Mahwah, NJ: Lawrence Erlbaum Associates.

Griffin, S., Case, R., & Siegler, R.S. (1994). Rightstart: Providing the central conceptual prerequisites for the first formal learning of arithmetic to students at risk for school failure. In K. McGilly (Ed.), *Classroom lessons: Integrating cognitive theory and classroom practice* (pp. 25–49). Cambridge, MA: MIT Press.

Gross-Tsur, V., Manor, O., & Shalev, R.S. (1996). Developmental dyscalculia: Prevalence and demographic features. *Developmental Medicine and Child Neurology, 38*, 25–33.

Gruber, O., Indefrey, P., Steinmetz, H., & Kleinschmidt, A. (2001). Dissociating neural correlates of cognitive components in mental calculation. *Cerebral Cortex, 11*, 350–359.

Guskey, T.R. (2008). The rest of the story: The power of formative classroom assessment depends on how you use the results. *Educational Leadership, 65*(4), 28–35.

Guttman, L.M. (2006). How student and parent goal orientations and classroom goal structures influence the math achievement of African Americans during the high school transition. *Contemporary Educational Psychology, 31*, 44–63.

Guttmann, E. (1937). Congenital arithmetic disability and acalculia (Henschen). *British Journal of Medical Psychology, 16*, 16–35.

Guy, S.C., Isquith, P.K., & Gioia, G.A. (2004). *Behavioral Rating Inventory of Executive Function—Self-Report Version (BRIEF-SR)*. Lutz, FL: Psychological Assessment Resources.

Habre, S. (2001). Visualization in multivariable calculus: The case of 3D-surfaces. *Focus on Learning Problems in Mathematics, 23*, 30–48.

Hale, J.B., & Fiorello, C.A. (2004). *School neuropsychology: A practitioner's handbook*. New York: Guilford Press.

Hale, J.B., Fiorello, C.A., Bertin, M., & Sherman, R. (2003). Predicting math achievement through neuropsychological interpretation of WISC-III variance. *Journal of Psychoeducational Assessment, 21*, 358–380.

Halpern, D.F. (1992). A cognitive approach to improving thinking skills in the sciences and mathematics. In D.F. Halpern (Ed.), *Enhancing thinking skills in the sciences and mathematics* (pp. 1–14). Mahwah, NJ: Lawrence Erlbaum Associates.

Halpern, D.F. (1998). Teaching critical thinking for transfer across domains. *American Psychologist, 53*, 449–455.

Halpern, D.F., Benbow, C.P., Geary, D.C., Gur, R.C., Hyde, J.S., & Gernsbacher, M.A. (2007). The science of sex differences in science and mathematics. *Psychological Science in the Public Interest, 8*, 1–51.

Hammill, D.D., Pearson, N.A., & Voress, J.K. (1993). *Developmental Test of Visual Perception—Second Edition (DTVP-2)*. Austin, TX: PRO-ED.

Hanson, S.A., & Hogan, T.P. (2000). Computational estimation skill of college students. *Journal for Research in Mathematics Education, 31*, 483–499.

Hardiman, P.T., & Mestre, J.P. (1989). Understanding multiplicative contexts involving fractions. *Journal of Educational Psychology, 81*, 547–557.

Harris, I.M., & Miniussi, C. (2003). Parietal lobe contribution to mental rotation demonstrated with rTMS. *Journal of Cognitive Neuroscience, 15*, 315–323.

Hasselbring, T.S., Goin, L.I., & Bransford, J.D. (1988, February). Developing automaticity in learning handicapped children: The role of computerized drill and practice. *Focus on Exceptional Children, 20*, 1–7.

Hatano, G., Amaiwa, S., & Inagaki, K. (1996). "Buggy algorithms" as attractive variants. *Journal of Mathematical Behavior, 15*, 285–302.

Haynes, J., Sakai, K., Rees, G., Gilbert, S., Frith, C., & Passingham, R. (2007). Reading hidden intentions in the human brain. *Current Biology, 17*, 323–328.

Hecht, S.A., Close, L., & Santisi, M. (2003). Sources of individual differences in fraction skills. *Journal of Experimental Child Psychology, 86*, 277–302.

Hecht, S.A., Torgesen, J.K., Wagner, R.K., & Rashotte, C.A. (2001). The relations between phonological processing abilities and emerging individual differences in mathematical computational skills: A longitudinal study from second to fifth grades. *Journal of Experimental Child Psychology, 79,* 192–227.

Hegarty, M. (1992). Mental animation: Inferring motion from static diagrams of mechanical systems. *Journal of Experimental Psychology: Learning, Memory, and Cognition, 18,* 1084–1102.

Hegarty, M. (2004). Mechanical reasoning by mental simulation. *Trends in Cognitive Sciences, 8,* 280–285.

Hegarty, M., & Kozhevnikov, M. (1999). Types of visual-spatial representations and mathematical problem solving. *Journal of Educational Psychology, 91,* 684–689.

Hembree, R. (1990). The nature, effects, and relief of mathematics anxiety. *Journal for Research in Mathematics Education, 21,* 33–46.

Hembree, R. (1992). Experimental and relational studies in problem solving: A meta-analysis. *Journal for Research in Mathematics Education, 23,* 242–273.

Henry, M.K. (2005). The history and structure of written English. In J.R. Birsh (Ed.), *Multisensory teaching of basic language skills* (2nd ed., pp. 151–170). Baltimore: Paul H. Brookes Publishing Co.

Heritage, M. (2007). Formative assessment: What do teachers need to know and do? *Phi Delta Kappan, 89,* 140–145.

Hermelin, B., & O'Connor, N. (1986). Spatial representations in mathematically and in artistically gifted children. *British Journal of Educational Psychology, 56,* 150–157.

Hermelin, B., & O'Connor, N. (1990). Factors and prime: A specific numerical ability. *Psychological Medicine, 20,* 163–169.

Hermer-Vazquez, L., Spelke, E., & Katsnelson, A. (1999). Source of flexibility in human cognition: Dual task studies of space and language. *Cognitive Psychology, 39,* 3–36.

Hespos, S.J., & Rochat, P. (1997). Dynamic representation in infancy. *Cognition, 64,* 153–189.

Hiebert, J., Gallimore, R., Garnier, H., Givvin, K.B., Hollingworth, H., Jacobs, J., et al. (2003). *Teaching mathematics in seven countries: Results from the TIMSS 1999 video study.* Washington, DC: U.S. Department of Education, National Center for Educational Statistics.

Hill, H.C., Ball, D.L., & Schilling, S.G. (2008). Unpacking pedagogical content knowledge: Conceptualizing and measuring teachers' topic specific knowledge of students. *Journal for Research in Mathematics Education, 39,* 372–400.

Hitch, G.J. (1978). The role of short-term working memory in mental arithmetic. *Cognitive Psychology, 10,* 302–323.

Hittmair-Delazer, M., Semenza, C., & Denes, G. (1994). Concepts and facts in calculation. *Brain, 117,* 715–728.

Hobbs, J. & Kreinovich, V. (2006). Optimal choice of granularity in commonsense estimation: Why half-orders of magnitude. *International Journal of Intelligent Systems, 21,* 843–855.

Hodges, J.R., Spatt, J., & Patterson, K. (1999). "What" and "how": Evidence for the dissociation of object knowledge and mechanical problem-solving skills in the human brain. *Proceedings of the National Academy of Sciences USA, 96,* 9444–9448.

Holloway, I., & Ansari, D. (2008). Domain-specific and domain-general changes in children's development of number comparison. *Developmental Science, 11,* 644–649.

Hoosain, R., & Salili, F. (1987). Language differences in pronunciation speed for numbers, digit span and mathematical ability. *Psychologia, 30,* 34–38.

Hubbard, E.M., Piazza, M., Pinel, P., & Dehaene, S. (2005). Interactions between number and space in parietal cortex. *Nature Reviews Neuroscience, 6,* 435–448.

Hudson, T. (1983). Correspondences and numerical differences between disjoint sets. *Child Development, 54,* 84–90.

Hugdahl, K., Rund, B.R., Lund, A., Asbjørnsen, A., Egeland, J., Ersland, L., et al. (2004). Brain activation measured with fMRI during a mental arithmetic task in schizophrenia and major depression. *American Journal of Psychiatry, 161,* 286–293.

Humphreys, L.G., Lubinski, D., & Yao, G. (1993). Utility of predicting group membership and the role of spatial visualization in becoming an engineer, physical scientist, or artist. *Journal of Applied Psychology, 78,* 250–261.

Huntley-Fenner, G.N., & Cannon, E. (2000). Preschoolers' magnitude comparisons are mediated by a preverbal analog magnitude mechanism. *Psychological Science, 11,* 147–152.

Huntley-Fenner, G., Carey, S., & Solimando, A. (2002). Objects are individuals but stuff doesn't count: Perceived rigidity and cohesiveness influence infants' representations of small groups of discrete entities. *Cognition, 85,* 203–221.

Hurewitz, F., Papafragou, A., Gleitman, L., & Gelman, R. (2006). Asymmetries in the acquisition of numbers and quantifiers. *Language Learning and Development, 2,* 77–96.

Hutton, L.A., & Levitt, E. (1987). An academic approach to the remediation of mathematics anxiety. In R. Schwarzer, H.M. vander Ploeg, & C.D. Spielberger (Eds.), *Advances in test anxiety research* (Vol. 5, pp. 207–211). Berwyn, PA: Swets North America.

IDEA 2004 Final Rule (2006, August 14). 34 CFR, Parts 300 and 301. *Federal Register, 71*(156). Retrieved on September 21, 2007, from http://idea.ed.gov/download/finalregulations.pdf

Individuals with Disabilities Education Improvement Act (IDEA) of 2004, PL 108-446, 20 U.S.C. §§ 1400 *et seq.*

Ischebeck, A., Zamarian, L., Egger, K., Schocke, M., & Delazer, M. (2007). Imaging early practice effects in arithmetic. *NeuroImage, 36,* 993–1003.

Ischebeck, A., Zamarian, L., Siedentopf, C., Koppelstätter, F., Benke, T., Felber, S., et al. (2006). How specifically do we learn? Imaging the learning of multiplication and subtraction. *NeuroImage, 30,* 1365–1375.

Ito, Y., & Hatta, T. (2004). Spatial structure of quantitative representation of numbers: Evidence from the SNARC effect. *Memory & Cognition, 32,* 662–673.

Iuculano, T., Tang, J., Hall, C.W.B., & Butterworth, B. (2008). Core information processing deficits in developmental dyscalculia and low numeracy. *Developmental Science, 11,* 669–680.

Iversen, W., Nuerk, H.-C., & Willmes, K. (2004). Do signers think differently? The processing of number parity in deaf participants. *Cortex, 40,* 176–178.

Jeong, Y., Levine, S.C., & Huttenlocher, J. (2007). The development of proportional reasoning: Effect of continuous v. discrete quantities. *Journal of Cognition and Development, 8,* 237–256.

Johnson, E.S. (1984). Sex differences in problem solving. *Journal of Educational Psychology, 76,* 1359–1371.

Jones, P.L. (1982). Learning mathematics in a second language: A problem with more and less. *Educational Studies in Mathematics, 13,* 269–287.

Jordan, K., Wüstenberg, T., Heinze, H.J., Peters, M. & Jancke, L. (2002). Women and men exhibit different cortical activation patterns during mental rotation tasks. *Neuropsychologia, 40,* 2397–2408.

Jordan, N.C., Hanich, L.B., & Kaplan, D. (2003a). A longitudinal study of mathematical competencies in children with specific mathematics difficulties versus children with co-morbid mathematics and reading difficulties. *Child Development, 74,* 834–850.

Jordan, N.C., Hanich, L.B., & Kaplan, D. (2003b). Arithmetic fact mastery in young children: A longitudinal investigation. *Journal of Experimental Child Psychology, 85,* 103–119.

Jordan, N.C., Huttenlocher, J., & Levine, S.C. (1992). Differential calculation abilities in young children from middle- and low-income families. *Developmental Psychology, 28,* 644–653.

Jordan, N.C., Kaplan, D., Locuniak, M.N., & Ramineni, C. (2007). Predicting first-grade math achievement from developmental number sense trajectories. *Learning Disabilities Research and Practice, 22,* 36–46.

Jordan, N.C., Kaplan, D., Olah, L., & Locuniak, M.N. (2006). Number sense growth in kindergarten: A longitudinal investigation of children at risk for mathematics difficulties. *Child Development, 77,* 153–175.

Jordan, N.C., Kaplan, D., Ramineni, C., & Locuniak, M.N. (2008). Development of number combination skill in the early school years: When do fingers help? *Developmental Science, 11,* 662–668.

Jordan, N.C., Levine, S.C., & Huttenlocher, J. (1995). Calculation abilities of young children with different patterns of cognitive functioning. *Journal of Learning Disabilities, 28,* 53–64.

Jordan, N.C., & Montani, T.O. (1997). Cognitive arithmetic and problem solving: A comparison of children with specific and general mathematics difficulties. *Journal of Learning Disabilities, 30,* 624–634, 684.

Kadosh, R.C., Kadosh, K.C., Kaas, A., Henik, A., & Goebel, R. (2007). Notation-dependent and -independent representations of numbers in the parietal lobes. *Neuron, 53,* 307–314.

Kaemingk, K.L., Carey, M.E., Moore, I.M., Herzer, M., & Hutter, J.J. (2004). Math weaknesses in survivors of acute lymphoblastic leukemia compared to healthy children. *Child Neuropsychology, 10,* 14–23.

Kalchman, M., Moss, J., & Case, R. (2001). Psychological models for the development of mathematical understanding: Rational numbers and functions. In S. Carver & D. Klahr (Eds.), *Cognition and instruction* (pp. 1–38). Mahwah, NJ: Lawrence Erlbaum Associates.

Kalyuga, S., Chandler, P., Tuovinen, J. & Sweller, J. (2001). When problem solving is superior to studying worked examples. *Journal of Educational Psychology, 93,* 579–588.

Kaminski, J.A., Sloutsky, V.M., & Heckler, A.F. (2008, April 5). Learning theory: The advantage of abstract examples in learning math. *Science, 320,* 454–455.

Kane, R. (1968). The readability of mathematical English. *Journal of Research in Science Teaching, 5,* 296–298.

Kaplan, R.F., & Stevens, M.C. (2002). A review of adult ADHD: A neuropsychological and neuroimaging perspective. *CNS Spectrums, 7,* 355–362.

Kaufmann, L., Koppelstaetter, F., Siedentopf, C., Haala, I., Haberlandt, E., Zimmerhackl, L.B., et al. (2006). Neural correlates of the number-size interference task in children. *NeuroReport, 17,* 587–591.

Kaufmann, L., & Nuerk, H.-C. (2008). Basic number processing deficits in ADHD: A broad examination of elementary and complex number processing skills in 9- to 12-year-old children with ADHD-C. *Developmental Science, 11,* 692–699.

Kelly, A.E., & Lesh, R.A. (2000). Preface. In A.E. Kelly & R.A. Lesh (Eds.), *Handbook of research design in mathematics and science education* (pp. 9–16). Mahwah, NJ: Lawrence Erlbaum Associates.

Kessel, C., & Ma, L. (2001). Mathematicians and the preparation of elementary teachers. In D. Holton (Ed.), *The teaching and learning of mathematics at university level* (pp. 467–480). Boston: Kluwer Academic.

Kessler, R.C., Adler, L., Barkley, R., Biederman, J., Conners, C.K., Demler, O., et al. (2006). The prevalence and correlates of adult ADHD in the United States: Results from the National Comorbidity Survey Replication. *American Journal of Psychiatry, 163,* 716–723.

Kiefer, M., Apel, A., & Weisbrod, M. (2002). Arithmetic fact retrieval and working memory in schizophrenia. *Schizophrenia Research, 53,* 219–227.

Kirshner, D., & Awtry, T. (2004). Visual salience of algebraic transformations. *Journal for Research in Mathematics Education, 35,* 224–257.

Klein, A.S., Beishuizen, M., & Treffers, A. (1998). The empty number line in Dutch second grades: Realistic versus gradual program design. *Journal for Research in Mathematics Education, 29,* 443–464.

Klein, J.S., & Bisanz, J. (2000). Preschoolers doing arithmetic: The concepts are willing but the working memory is weak. *Canadian Journal of Experimental Psychology, 54,* 105–116.

Klingberg, T., Forssberg, H., & Westerberg, H. (2002). Increased brain activity in frontal and parietal cortex underlies the development of visuospatial working memory capacity during childhood. *Journal of Cognitive Neuroscience, 14,* 1–10.

Knauff, M., & Johnson-Laird, P.N. (2002). Visual imagery can impede reasoning. *Memory & Cognition, 30,* 363–371.

Knauff, M., Mulack, T., Kassubek, J., Salih, H., & Greenlee, M.W. (2002). Spatial imagery in deductive reasoning: A functional MRI study. *Cognitive Brain Research, 13,* 203–212.

Knuth, E.J., Stephans, A.C., McNeil, N.M., & Alibali, M.W. (2006). Does understanding the equal sign matter? Evidence from solving equations. *Journal for Research in Mathematics Education, 37,* 297–312.

Kobayashi, T., Hiraki, K., & Hasegawa, T. (2005). Auditory-visual intermodal matching of small numerosities in 6-month-old infants. *Developmental Science, 8,* 409–419.

Koedinger, K.R., Alibali, M.W., & Nathan, M.J. (2008). Trade-offs between grounded and abstract representations: Evidence from algebra problem solving. *Cognitive Science, 32,* 366–397.

Kong, J., Wang, C., Kwong, K., Vangel, M., Chua, E., & Gollub, R. (2005). The neural substrate of arithmetic operations and procedure complexity. *Cognitive Brain Research, 22,* 397–405.

Kopera-Frye, K., Dehaene, S., & Streissguth, A.P. (1996). Impairments of number processing induced by prenatal alcohol exposure. *Neuropsychologia, 34,* 1187–1196.

Kosslyn, S.M. (1994). *Image and brain: The resolution of the imagery debate.* Cambridge, MA: MIT Press.

Kouider, S., Halberda, J., Wood, J.N., & Carey, S. (2006). Acquisition of English number marking: The singular-plural distinction. *Language Learning and Development, 2,* 1–25.

Kovacs, M. (1992). *Children's Depression Inventory (CDI).* North Tonawanda, NY: Multi-Health Systems.

Kovas, Y., Haworth, C.M.A., Dale, P.S., & Plomin, R. (2007). The genetic and environmental origins of learning abilities and disabilities in the early school years. *Monographs of the Society for Research in Child Development, 72*(3, Serial No. 288).

Kovas, Y., Petrill, S.A., & Plomin, R. (2007). The origins of diverse domains of mathematics: Generalist genes but specialist environments. *Journal of Educational Psychology, 99,* 128–139.

Kozhevnikov, M., Hegarty, M., & Mayer, R.E. (2002). Revising the visualizer/verbalizer dimension: Evidence for two types of visualizers. *Cognition and Instruction, 20,* 47–77.

Kozhevnikov, M., Kosslyn, S., & Shephard, J. (2005). Spatial versus object visualizers: A new characterization of visual cognitive style. *Memory & Cognition, 33,* 710–726.

Kroesbergen, E.H., Van Luit, J.E.H., & Maas, C.J.M. (2004). Effectiveness of explicit and constructivist mathematics instruction for low-achieving students in the Netherlands. *The Elementary School Journal, 104,* 233–251.

Kroger, J., Cohen, J., & Johnson-Laird, P.N. (2001). *A double dissociation between logic and mathematics: A functional magnetic resonance imaging study.* Unpublished manuscript.

Kucian, K., Loenneker, T., Dietrich, T., Martin, E., & von Aster, M. (2005). Gender differences in brain activation patterns during mental rotation and number related cognitive tasks. *Psychology Science, 47,* 112–131.

Kulp, M.T. (1999). Relationship between visual motor integration skill and academic performance in kindergarten through third grade. *Optometry and Vision Science, 76,* 159–163.

Kulp, M.T., Earley, M.J., Mitchell, G.L., Timmerman, L.M., Frasco, C.S., & Geiger, M.E. (2004). Are visual perceptual skills related to mathematics ability in second through sixth grade children? *Focus on Learning Problems in Mathematics, 26,* 44–51.

Kulp, M.T., & Schmidt, P.P. (1996). Effect of oculomotor and other visual skills on reading performance: A literature review. *Optometry and Vision Science, 73,* 283–292.

Kwon, H., Reiss, A.L., & Menon, V. (2002). Neural basis of protracted developmental changes in visuo-spatial working memory. *Proceedings of the National Academy of Sciences USA, 99,* 13336–13341.

Lachmann, T. (2002). Reading disability as a deficit in functional coordination. In E. Witruk, A.D. Friederici, & T. Lachmann (Eds.), *Basic functions of language, reading, and reading disability* (pp. 165–198). Boston: Kluwer Academic.

Lagace, D.C., Kutcher, S.P., & Robertson, H.A. (2003). Mathematics deficits in adolescents with bipolar I disorder. *American Journal of Psychiatry, 160,* 100–104.

Landerl, K., Bevan, A. & Butterworth, B. (2004). Developmental dyscalculia and basic numerical capacities: A study of 8-9-year-old students. *Cognition, 93,* 99–125.

Laski, E.V., & Siegler, R.S. (2007). Is 27 a big number? Correlational and causal connections among numerical categorization, number line estimation, and numerical magnitude comparison. *Child Development, 76,* 1723–1743.

Lee, K., Lim, Z.Y., Yeong, S.H.M., Ng, S.F., Venkatraman, V., & Chee, M.W.L. (2007). Strategic differences in algebraic problem solving: Neuroanatomical correlates. *Brain Research, 1155,* 163–171.

Lee, K.M. (2000). Cortical areas differentially involved in multiplication and subtraction: A functional magnetic resonance imaging study and correlation with a case of selective acalculia. *Annals of Neurology, 48,* 657–661.

Lee, K.M., & Kang, S.Y. (2002). Arithmetic operation and working memory: Differential suppression in dual tasks. *Cognition, 83,* B63–B68.

Lee, S.A., & Spelke, E.S. (2008). Children's use of geometry for reorientation. *Developmental Science, 11,* 743–749.

LeFevre, J.-A., Bisanz, J., Daley, K.E., Buffone, L., Greenham, S.L., & Sadesky, G.S. (1996). Multiple routes to solution of single-digit multiplication problems. *Journal of Experimental Psychology: General, 125,* 284–306.

LeFevre, J.-A., Bisanz, J., & Mrkonjic, L. (1988). Cognitive arithmetic: Evidence for obligatory activation of arithmetic facts. *Memory & Cognition, 16,* 45–53.

LeFevre, J.-A., Greenham, S.L., & Waheed, N. (1993). The development of procedural and conceptual knowledge in computational estimation. *Cognition and Instruction, 11,* 95–132.

LeFevre, J.-A., & Liu, J. (1997). The role of experience in numerical skill: Multiplication performance in adults from Canada and China. *Mathematical Cognition, 3,* 31–62.

LeFevre, J.-A., Smith-Chant, B.L., Fast, L., Skwarchuk, S.-L., Sargla, E., Arnup, J.S., et al. (2006). What counts as knowing? The development of conceptual and procedural knowledge of counting from kindergarten through grade 2. *Journal of Experimental Child Psychology, 93,* 285–303.

Lehmann, W., & Jüling, I. (2002). Spatial reasoning and mathematical abilities: Independent constructs or two sides of the same coin? *Psychologie in Erziehung und Unterricht, 49,* 31–43.

Lehto, J. (1995). Working memory and school achievement in the ninth form. *Educational Psychology, 15,* 271–281.

Lemaire, P., Barrett, S.E., Fayol, M., & Abdi, H. (1994). Automatic activation of addition and multiplication facts in elementary school children. *Journal of Experimental Child Psychology, 57,* 224–258.

Lemaire, P., & Lecacheur, M. (2002). Children's strategies in computational estimation. *Journal of Experimental Child Psychology, 82,* 281–304.

Lemaire, P., & Siegler, R.S. (1995). Four aspects of strategic change: Contributions to children's learning of multiplication. *Journal of Experimental Psychology: General, 124,* 83–97.

Lemer, C., Dehaene, S., Spelke, E., & Cohen, L. (2003). Approximate quantities and exact number words: Dissociable systems. *Neuropsychologia, 41,* 1942–1958.

Leng, X., & Shaw, G.L. (1991). Toward a neural theory of higher brain function using music as a window. *Concepts in Neuroscience, 2,* 229–258.

Levin, H.S., Mattis, S., Ruff, R.M., Eisenberg, H.M., Marshall, L.F., Tabaddor, K., et al. (1987). Neurobehavioral outcome following minor head injury: A three-center study. *Journal of Neurosurgery, 66,* 234–243.

Levine, S., Huttenlocher, J., Taylor, A., & Langrock, A. (1999). Early sex differences in spatial skill. *Developmental Psychology, 35,* 940–949.

Levy, L.M., Reis, I.L., & Grafman, J. (1999). Metabolic abnormalities detected by 1H-MRS in dyscalculia and dysgraphia. *Neurology, 53,* 639–641.

Lewis, A.B. (1989). Training students to represent arithmetic word problems. *Journal of Educational Psychology, 81,* 521–531.

Lewis, A.B., & Mayer, R.E. (1987). Students' miscomprehension of relational statements in arithmetic word problems. *Journal of Educational Psychology, 79,* 363–371.

Lewis, C., Hitch, G.J., & Walker, P. (1994). The prevalence of specific arithmetic difficulties and specific reading difficulties in 9 and 10 year old boys and girls. *Journal of Child Psychology and Psychiatry, 35,* 283–292.

Lichtenstein, R., & Klotz, M.B. (2007, November). Deciphering the federal regulations on identifying children with specific learning disabilities. *NASP Communiqué, 36.* Retrieved on October 16, 2008, from http://www.nasponline.org/publications/cq/mocq363regs.aspx

Lim, B. (2001). Relationships between linguistics and understanding fractions by monolingual English speaking students and bilingual Korean American students. *Focus on Learning Problems in Mathematics, 23,* 85–101.

Lim, T.K. (1994). Gender-related differences in intelligence: Application of confirmatory factor analysis. *Intelligence, 19,* 179–192.

Lindsay, R.L., Tomazic, T., Levine, M.D., & Accardo, P.J. (2001). Attentional function as measured by a Continuous Performance Task in children with dyscalculia. *Journal of Developmental and Behavioral Pediatrics, 22,* 287–292.

Linn, M.C., Layman, J., & Nachmias, R. (1987). Cognitive consequences of microcomputer-based laboratories: Graphing skills development. *Contemporary Educational Psychology, 12,* 244–253.

Linn, M.C., & Petersen, A.C. (1985). Emergence and characterization of sex differences in spatial ability: A meta-analysis. *Child Development, 56,* 1479–1498.

Linnebrink, E. (2005). The dilemma of performance-approach goals: The use of multiple goal contexts to promote students' motivation and learning. *Journal of Educational Psychology, 97,* 197–213.

Lipton, J.S., & Spelke, E.S. (2006). Preschool children master the logic of the verbal counting routine. *Cognition, 98,* B57–B66.

Liu, L.L., Uttal, D.H., Marulis, L.M., & Newcombe, N.S. (2008, May). Training spatial skills: What works, for whom, why and for how long? Poster session presented at the 20th annual meeting of the Association for Psychological Science, Chicago, IL.

Locke, J. (1690/1824). *An essay concerning human understanding* (Book II, Chapt. XVI, §1). Retrieved March 20, 2006, from http://oll.libertyfund.org/Texts/Locke0154/Works/0128-01_Bk.html

Loney, J., Kramer, J., & Millich, R. (1979). *The hyperkinetic child grows up: Predictors of symptoms, delinquency, and achievement at follow-up.* Paper presented at the Annual Meeting of the American Association for the Advancement of Science, Houston, TX.

Low, R., & Over, R. (1993). Gender differences in solution of algebraic word problems containing irrelevant information. *Journal of Educational Psychology, 85,* 331–339.

Low, R., Over, R., Doolan, L., & Michell, S. (1994). Solution of algebraic word problems following training in identifying necessary and sufficient information within problems. *American Journal of Psychology, 107,* 423–439.

Lowrie, T. (2000). A case of an individual's reluctance to visualize. *Focus on Learning Problems in Mathematics, 22,* 17–26.

Luchins, A.S. (1942). Mechanization in problem solving. *Psychological Monographs, 54*(6, Serial No. 95).

Luciana, M. (2003). The neural and functional development of human prefrontal cortex. In M. de Haan & M.H. Johnson (Eds.), *The cognitive neuroscience of development* (pp. 157–179). Brighton, NY: Psychology Press.

Luckner, J.L., & McNeill, J.H. (1994). Performance of a group of deaf and hard-of-hearing students and a comparison group of hearing students on a series of problem-solving tasks. *American Annals of the Deaf, 139,* 371–376.

Lundberg, I., & Sterner, G. (2006). Reading, arithmetic, and task orientation—How are they related? *Annals of Dyslexia, 56,* 361–377.

Ma, L.-P. (1999). *The knowing and teaching of elementary mathematics.* Mahwah, NJ: Lawrence Erlbaum Associates.

Mackintosh, N.J. (1998). *IQ and human intelligence.* New York: Oxford University Press.

Mahone, E.M., Hagelthorn, K.M., Cutting, L.E., Schuerholz, L.J., Pelletier, S.F., Rawlins, C., et al. (2002). Effects of IQ on executive function measures in children with ADHD. *Child Neuropsychology, 8,* 52–65.

Manalo, E., Bunnell, J.K., & Stillman, J.A. (2000). The use of process mnemonics in teaching students with mathematics learning disabilities. *Learning Disability Quarterly, 23,* 137–156.

Manor, O., Shalev, R.S., Joseph, A., & Gross-Tsur, V. (2000). Arithmetic skills in kindergarten children with developmental language disorders. *European Journal of Paediatric Neurology, 5,* 71–77.

Marshall, J.C., & Fink, G.R. (2001). Spatial cognition: Where we were and where we are. *NeuroImage, 14,* S2–S7.

Marzocchi, G.M., Lucangeli, D., DeMeo, T., Fini, F., & Cornoldi, C. (2002). The disturbing effect of irrelevant information on arithmetic problem solving in inattentive children. *Developmental Neuropsychology, 21,* 73–92.

Masters, M.S., & Sanders, B. (1993). Is the gender difference in mental rotation disappearing? *Behavior Genetics, 23,* 337–341.

Mathewson, J.H. (1999). Visual-spatial thinking: An aspect of science overlooked by educators. *Science Education, 83,* 33–54.

Mayfield, K.H., & Chase, P.N. (2002). The effects of cumulative practice on mathematics problem solving. *Journal of Applied Behavior Analysis, 35,* 105–123.

Mazzocco, M.M.M. (2005). Challenges in identifying target skills for math disability screening and intervention. *Journal of Learning Disabilities, 38,* 318–323.

Mazzocco, M.M.M., & Devlin, K.T. (2008). Parts and 'holes': Gaps in rational number sense among children with and without mathematical learning disabilities. *Developmental Science, 11,* 681–691.

Mazzocco, M.M.M., & Myers, G.F. (2003). Complexities in identifying and defining mathematics learning disability in the primary school-age years. *Annals of Dyslexia, 53,* 218–253.

Mazzocco, M.M.M., & Thompson, R. (2005). Kindergarten predictors of math learning disability. *Learning Disabilities Research and Practice, 20,* 142–155.

McCloskey, M., & Caramazza, A. (1987). Cognitive mechanisms in normal and impaired number processing. In G. Deloche & X. Seron (Eds.), *Mathematical disabilities: A cognitive neuropsychological perspective* (pp. 201–219). Mahwah, NJ: Lawrence Erlbaum Associates.

McCloskey, M., Caramazza, A., & Basili, A. (1985). Cognitive mechanisms in number processing and calculation: Evidence from dyscalculia. *Brain and Cognition, 4,* 171–196.

McComb, K., Parker, C., & Pusey, A. (1994). Roaring and numerical assessment in contests between groups of female lions, *Panthera leo. Animal Behaviour, 47,* 379–387.

McCue, M., Goldstein, G., Shelly, C., & Katz, L. (1986). Cognitive profiles of some subtypes of learning disabled adults. *Archives of Clinical Neuropsychology, 1,* 13–23.

McDermott, L.C., Rosenquist, M.L., & van Zee, E.H. (1987). Student difficulties in connecting graphs and physics: Examples from kinematics. *American Journal of Physics, 55,* 503–513.

McGuinness, D. (1993). Sex differences in cognitive style: Implications for mathematics performance and achievement. In L.A. Penner, G.M. Batsche, H.M. Knoff, & D.L. Nelson (Eds.), *The challenge of mathematics and science education: Psychology's response* (pp. 251–274). Washington, DC: American Psychological Association.

McLeod, D.B. (1992). Research on affect in mathematics education: A reconceptualization. In D.A. Grouws & D.B. McLeod (Eds.), *Handbook of research on mathematics teaching and learning* (pp. 575–596). New York: Macmillan.

McLeod, T., & Armstrong, S. (1982). Learning disabilities in mathematics-skill deficits and remedial approaches at the intermediate and secondary grades. *Learning Disability Quarterly, 5,* 305–311.

McMonnies, C.W. (1992). Visuo-spatial discrimination and mirror image letter reversals in reading. *Journal of the American Optometric Association, 63,* 698–704.

McNeil, N. (2007). U-shaped development in math: 7-year-olds outperform 9-year-olds on equivalence problems. *Developmental Psychology, 43,* 687–695.

McNeil, N.M., & Alibali, M.W. (2005). Knowledge change as a function of mathematics experience: All contexts are not created equal. *Journal of Cognition and Development, 6,* 285–306.

McNeil, N.M., Grandau, L., Knuth, E.J., Alibali, M.W., Stephans, A.C., Hattikudur, S., et al. (2006). Middle-school students' understanding of the equal sign: The books they read can't help. *Cognition and Instruction, 24,* 367–385.

Meece, J.L., Anderman, E.M., & Anderman, L.H. (2006). Classroom goal structure, student motivation, and academic achievement. *Annual Review of Psychology, 57,* 487–503.

Meehan, A.M. (1984). A meta-analysis of sex differences in formal operational thought. *Child Development, 55,* 1110–1124.

Menon, V., Rivera, S.M., White, C.D., Glover, G.H., & Reiss, A.L. (2000). Dissociating prefrontal and parietal cortex activation during arithmetic processing. *NeuroImage, 12,* 357–365.

Mestre, J.P. (1988). The role of language comprehension in mathematics and problem solving. In R.R. Cocking & J.P. Mestre (Eds.), *Linguistic and cultural influences on learning mathematics* (pp. 201–220). Mahwah, NJ: Lawrence Erlbaum Associates.

Meyler, A., Keller, T.A., Cherkassky, V.L., Gabrieli, J.D.E., & Just, M.A. (2008). Modifying the brain activation of poor readers during sentence comprehension with extended remedial instruction: A longitudinal study of neuroplasticity. *Neuropsychologia, 46,* 2580–2592.

Miles, T.R., Haslum, M.N., & Wheeler, T.J. (2001). The mathematical abilities of dyslexic 10-year-olds. *Annals of Dyslexia, 51,* 299–321.

Miller, K.F. (1984). Child as measurer of all things: Measurement procedures and the development of quantitative concepts. In C. Sophian (Ed.), *Origins of cognitive skills* (pp. 193–228). Mahwah, NJ: Lawrence Erlbaum Associates.

Miller, K.F., Kelly, M., & Zhou, X. (2005). Learning mathematics in China and the United States: Cross-cultural insights into the nature and course of preschool mathematical development. In J.I.D. Campbell (Ed.), *Handbook of mathematical cognition* (pp. 163–178). New York: Psychology Press.

Miller, K.F., Major, S.M., Shu, H., & Zhang, H. (2000). Ordinal knowledge: Number names and number concepts in Chinese and English. *Canadian Journal of Experimental Psychology, 54*, 129–140.

Miller, K.F., & Paredes, D.R. (1990). Starting to add worse: Effects of learning to multiply on children's addition. *Cognition, 37*, 213–242.

Miller, K., Smith, C.M., Zhu, J., & Zhang, H. (1995). Preschool origins of cross-national differences in mathematical competence: The role of number-naming systems. *Psychological Science, 6*, 56–60.

Mills, C.J., Ablard, K.E., & Stumpf, H. (1993). Gender differences in academically talented young students' mathematical reasoning: Patterns across age and subskills. *Journal of Educational Psychology, 85*, 340–346.

Miura, I.T., Okamoto, Y., Vlahovic-Stetic, V., Kim, C.C., & Han, J.H. (1999). Language supports for children's understanding of numerical fractions: Cross-national comparisons. *Journal of Experimental Child Psychology, 74*, 356–365.

Mix K.S. (1999). Similarity and numerical equivalence: Appearances count. *Cognitive Development, 14*, 269–297.

Mix, K.S. (2002). The construction of number concepts. *Cognitive Development, 17*, 1345–1363.

Mix, K.S., Huttenlocher, J., & Levine, S.C. (2002a). Multiple cues for quantification in infancy: Is number one of them? *Psychological Bulletin, 128*, 278–294.

Mix, K.S., Huttenlocher, J., & Levine, S.C. (2002b). *Quantitative development in infancy and early childhood.* New York: Oxford University Press.

Mix, K.S., Sandhofer, C.M., & Baroody, A.J. (2005). Number words and number concepts: The interplay of verbal and nonverbal quantification in early childhood. *Advances in Child Development and Behavior, 33*, 305–346.

Miyaki, A., Friedman, N.P., Emerson, M.J., Witzki, A.H., Howerter, A., & Wagner, T.D. (2000). The unity and diversity of executive functions and their contributions to complex frontal lobe tasks: A latent variable analysis. *Cognitive Psychology, 41*, 49–100.

Moats, L.C., & Farrell, M.L. (2005). Multisensory structured language education. In J.R. Birsh (Ed.), *Multisensory teaching of basic language skills* (2nd ed., pp. 23–41). Baltimore: Paul H. Brookes Publishing Co.

Moffat, S., Hampson, E., & Hatzipantelis, M. (1998). Navigation in a "virtual" maze: Sex differences and correlation with psychometric measures of spatial ability in humans. *Evolution and Human Behavior, 19*, 73–87.

Mokros, J.R., & Tinker, R.F. (1987). The impact of microcomputer-based labs on children's ability to interpret graphs. *Journal of Research in Science Teaching, 24*, 369–383.

Molko, N., Cachia, A., Rivière, D., Mangin, J.-F., Bruandet, M., LeBihan, D., et al. (2003). Functional and structural alterations of the intraparietal sulcus in a developmental dyscalculia of genetic origin. *Neuron, 40*, 847–858.

Morgan, A.E., Hynd, G.W., Riccio, C.A., & Hall, J. (1996). Validity of *DSM-IV* ADHD predominantly inattentive and combined types: Relationship to previous *DSM* diagnoses/subtype differences. *Journal of the American Academy of Child and Adolescent Psychiatry, 35*, 325–333.

Mousavi, S.Y., Low, R., & Sweller, J. (1995). Reducing cognitive load by mixing auditory and visual presentation modes. *Journal of Educational Psychology, 87*, 319–334.

Moyer, J.C., Sowder, L., Threadgill-Sowder, J., & Moyer, M.B. (1984). Story problem formats: Drawn versus verbal versus telegraphic. *Journal for Research in Mathematics Education, 15*, 342–351.

Moyer, R.S., & Landauer, T. (1967). Time required for judgments of numerical inequality. *Nature, 215*, 1519–1520.

Munn, P. (1998). Number symbols and symbolic function in preschoolers. In C. Donlan (Ed.), *The development of mathematical skills* (pp. 47–71). Hove, East Sussex, United Kingdom: Psychology Press.

Munro, J. (1979). Language abilities and math performance. *Reading Teacher, 32,* 900–915.

Murphy, M.M., Mazzocco, M.M.M., Hanich, L.B., & Early, M.C. (2007). Cognitive characteristics of children with mathematics learning disability (MLD) vary as a function of the cut-off criterion used to define MLD. *Journal of Learning Disabilities, 40,* 458–478.

National Council of Teachers of Mathematics. (1980). *An agenda for action.* Reston, VA: Author.

National Council of Teachers of Mathematics. (1989). *Curriculum and evaluation standards for school mathematics.* Reston, VA: Author.

National Council of Teachers of Mathematics. (1991). *Professional standards for teaching mathematics.* Reston, VA: Author.

National Council of Teachers of Mathematics. (1995). *Assessment standards for school mathematics.* Reston, VA: Author.

National Council of Teachers of Mathematics. (2000). *Principles and standards for school mathematics* (Overview). Reston, VA: Author.

National Council of Teachers of Mathematics. (2006). *Curriculum focal points: A quest for coherence.* Reston, VA: Author.

National Mathematics Advisory Panel. (2008). *Foundations for success: The final report of the National Mathematics Advisory Panel.* Washington, DC: U.S. Department of Education.

Newcombe, N.S., & Huttenlocher, J. (2000). *Making space.* Cambridge, MA: MIT Press.

Newcombe, N., Huttenlocher, J., & Learmonth, A. (1999). Infants' coding of location in continuous space. *Infant Behavior and Development, 22,* 483–510.

Newcombe, N.S., Mathason, L., & Terlecki, M. (2002). Maximization of spatial competence: More important than finding the cause of sex differences. In A. McGillicuddy-DeLisi & R. DeLisi (Eds.), *Biology, society, and behavior: The development of sex differences in cognition* (pp. 183–206). Westport, CT: Ablex.

Newcombe, N., Sluzenski, J., & Huttenlocher, J. (2005). Preexisting knowledge versus online learning. *Psychological Science, 16,* 222–227.

Newton, X. (2007). Reflections on math reforms in the U.S.: A cross-national perspective. *Phi Delta Kappan, 88,* 681–685.

No Child Left Behind Act of 2001, PL 107-110, 115 Stat. 1425, 20 U.S.C. §§ 6301, *et seq.*

Noël, M.P. (2005). Finger agnosia: A predictor of numerical abilities in children? *Child Neuropsychology, 11,* 413–430.

Noonan, J. (1990). Readability problems presented by mathematics text. *Early Child Development and Care, 54,* 57–81.

Novick, L.R. (1990). Representational transfer in problem solving. *Psychological Science, 1,* 128–132.

Nuerk, H.-C., Kaufmann, L., Zoppoth, S., & Willmes, K. (2004). On the development of the mental number line: More, less, or never holistic with increasing age? *Developmental Psychology, 40,* 1199–1211.

Nunes, T., & Moreno, C. (1998). Is hearing impairment a cause of difficulties in learning mathematics? In C. Donlan (Ed.), *The development of mathematical skills* (pp. 227–254). Hove, East Sussex, United Kingdom: Psychology Press.

Nuttall, R.L., Casey, M.B., & Pezaris, E. (2005). Spatial ability as a mediator of gender differences on mathematics tests: A biological-environmental framework. In A.M. Gallagher & J.C. Kaufman (Eds.), *Gender differences in mathematics* (pp. 121–142). New York: Cambridge University Press.

O'Boyle, M.W., Benbow, C.P., & Alexander, J.E. (1995). Sex differences, hemispheric laterality and associated brain activity in the intellectually gifted. *Developmental Neuropsychology, 11,* 415–443.

O'Boyle, M.W., Cunnington, R., Silk, T., Vaughn, D., Jackson, G., Syngeniotis, A., et al. (2005). Mathematically gifted male adolescents activate a unique brain network during mental rotation. *Cognitive Brain Research, 25,* 583–587.

O'Brien, T.C. (1973). Logical thinking in college students. *Educational Studies in Mathematics, 5,* 71–79.

O'Brien, T.C., Shapiro, B.J., & Reali, N.C. (1971). Logical thinking—Language and context. *Educational Studies in Mathematics, 4,* 201–219.

Opfer, J.E., & Siegler, R.S. (2007). Representational change and children's numerical estimation. *Cognitive Psychology, 55,* 169–195.

Opfer, J.E., & Thompson, C.A. (2006). Even early representations of numerical magnitude are spatially organized: Evidence for a directional magnitude bias in pre-reading preschoolers. In R. Sun & N. Miyake (Eds.), *Proceedings of the 28th Annual Cognitive Science Society* (pp. 639–644). Mahwah, NJ: Lawrence Erlbaum Associates.

Opfer, J.E., & Thompson, C.A. (2008). The trouble with transfer: Insights from microgenetic changes in the representation of numerical magnitude. *Child Development, 79,* 788–804.

Opfer, J.E., Thompson, C.A., & DeVries, J.M. (2007). Why children make "better" estimates of fractional magnitude than adults. In D.S. McNamara & J.G. Trafton (Eds.), *Proceedings of the 29th Annual Cognitive Science Society* (pp. 1361–1366). Nashville, TN: Cognitive Science Society.

Opfer, J.E., Young, C., & Krasa, N. (2008). [The relation between number line skill, spatial ability, and sex in the early grades.] Unpublished raw data.

Organisation for Economic Co-Operation and Development. (2007). *Programme for International Student Assessment (PISA). PISA 2006 results.* Retrieved March 5, 2008, from http://www.pisa.oecd.org/document/2/0,3343,en_32252351_32236191_39718850_1_1_1_1,00.html

Örnkloo, H. (2007). *Fitting objects into holes: On the development of spatial cognition skills. Acta Universitatis Upsaliensis.* Retrieved on June 1, 2008, from http://urn.kb.se/resolve?urn=urn:nbn:se:uu:diva-8316

Orr, E.W. (1987). *Twice as less: Black English and the performance of Black students in mathematics and science.* New York: W.W. Norton.

Ostad, S.A. (2000). Cognitive subtraction in a developmental perspective: Accuracy, speed-of-processing and strategy-use differences in normal and mathematically disabled children. *Focus on Learning Problems in Mathematics, 22,* 18–31.

Owen, K., & Lynn, R. (1993). Sex differences in primary cognitive abilities among blacks, Indians, and whites in South Africa. *Journal of Biosocial Science, 25,* 557–560.

Paik. J.H., & Mix, K.S. (2003). U.S. and Korean children's comprehension of fraction names: A reexamination of cross-national differences. *Child Development, 74,* 144–154.

Pajares, F. (2005). Gender differences in mathematics self-efficacy beliefs. In A.M. Gallagher & J.C. Kaufman (Eds.), *Gender differences in mathematics: An integrative psychological approach* (pp. 294–315). Cambridge, United Kingdom: Cambridge University Press.

Parmar, R.S., Cawley, J.F., & Frazita, R.R. (1996). Word problem-solving by students with and without mild disabilities. *Exceptional Children, 62,* 415–429.

Pashler, H., Rohrer, D., Cepeda, N.J., & Carpenter, S.K. (2007). Enhancing learning and retarding forgetting: Choices and consequences. *Psychonomic Bulletin & Review, 14,* 187–193.

Passolunghi, M.C., & Siegel, L.S. (2004). Working memory and access to numerical information in children with disability in mathematics. *Journal of Experimental Child Psychology, 88,* 348–367.

Perera, K. (1980). The assessment of linguistic difficulty in reading material. *Educational Research, 32,* 151–161.

Perrucci, V., Agnoli, F., & Albiero, P. (2003, April). *Stimulus type, age, and gender effects on mental rotation.* Poster presented at the Society for Research in Child Development Biennial Meeting, Tampa, FL.

Perrucci, V., Agnoli, F., & Albiero, P. (2008). Children's performance in mental rotation tasks: Orientation-free features flatten the slope. *Developmental Science, 11,* 732–742.

Petitto, A.L. (1990). Development of number line and measurement concepts. *Cognition and Instruction, 7,* 55–78.

Petrill, S.A. (2002). The case for general intelligence: A behavioral genetic perspective. In R.J. Sternberg & E.L. Grigorenko (Eds.), *The general factor of intelligence: How general is it?* (pp. 281–298). Mahwah, NJ: Lawrence Erlbaum Associates.

Pezaris, E., & Casey, M.B. (1991). Girls who use "masculine" problem-solving strategies on a spatial task: Proposed genetic and environmental factors. *Brain and Cognition, 17,* 1–22.

Phillips, L.H. (1997). Do "frontal tests" measure executive function? Issues of assessment and evidence from fluency tests. In P.M.A. Rabbit (Ed.), *Methodology of frontal and executive function* (pp. 191–214). Hove, East Sussex, United Kingdom: Psychology Press.

Piaget, J. (1975). Comments on mathematical education. *Contemporary Education, 47,* 5–10.

Piazza, M., Pinel, P., LeBihan, D., & Dehaene, S. (2007). A magnitude code common to numerosities and number symbols in human parietal cortex. *Neuron, 53,* 293–305.

Piburn, M.D., Reynolds, S.J., McAuliffe, C., Leedy, D.E., Birk, J.P., & Johnson, J.K. (2005). The role of visualization in learning from computer-based images. *International Journal of Science Education, 27,* 513–527.

Pimm, D. (1987). *Speaking mathematically: Communication in mathematics classrooms.* New York: Routledge.

Pliszka, S.R., Glahn, D.C., Semrud-Clikeman, M., Franklin, C., Perez, R., III, Xiong, J., et al. (2006). Neuroimaging of inhibitory control areas in children with Attention Deficit Hyperactivity Disorder who were treatment naive or in long-term treatment. *American Journal of Psychiatry, 163,* 1052–1060.

Plomin, R., Kovas, Y., & Haworth, C.M.A. (2007). Generalist genes: Genetic links between brain, mind, and education. *Mind, Brain, and Education, 1,* 11–19.

Polich, J.M., & Schwartz, S.H. (1974). The effect of problem size on representation in deductive problem solving. *Memory & Cognition, 2,* 683–686.

Polk, T.A., & Farah, M.J. (1995). Late experience alters vision. *Nature, 376,* 648–649.

Polk, T.A., Stallcup, M., Aguirre, G.K., Alsop, D.C., D'Esposito, M., Detre, J.A., et al. (2002). Neural specialization for letter recognition. *Journal of Cognitive Neuroscience, 14,* 145–159.

Pólya, G. (1945). *How to solve it.* Princeton, NJ: Princeton University Press.

Postma, A., Izendoorn, R., & DeHaan, E.H.F. (1998). Sex differences in object location memory. *Brain and Cognition, 36,* 334–345.

Power, R. & DalMartello, M.F. (1997). From 834 to eight thirty four: The reading of Arabic numerals by seven-year-old children. *Mathematical Cognition, 3,* 63–85.

Prabhakaran, V., Rypma, B., & Gabrieli, J.D.E. (2001). Neural substrates of mathematical reasoning: A functional magnetic resonance imaging study of neocortical activation during performance of the necessary arithmetic operations test. *Neuropsychology, 15,* 115–127.

Presmeg, N. (1986). Visualization and mathematical giftedness. *Educational Studies in Mathematics, 17,* 297–311.

Presmeg, N. (1992). Prototypes, metonymies and imaginative rationality in high school mathematics. *Educational Studies in Mathematics, 23,* 595–610.

Price, G.R., Holloway, I., Räsänen, P., Vesterinen, M., & Ansari, D. (2007). Impaired parietal magnitude processing in developmental dyscalculia. *Current Biology, 17,* R1042–R1043.

Qin, Y., Carter, C.S., Silk, E.M., Stenger, V.A., Fissell, K., Goode, A., et al. (2004). The change of the brain activation patterns as children learn algebra equation solving. *Proceedings of the National Academy of Sciences USA, 101,* 5686–5691.

Quinn, P.C., Slater, A.M., Brown, E., & Hayes, R.A. (2001). Developmental change in form categorization in early infancy. *British Journal of Developmental Psychology, 19,* 207–218.

Ramani, G.B., & Siegler, R.S. (2008). Promoting broad and stable improvements in low-income children's numerical knowledge through playing number board games. *Child Development, 79,* 375–394.

Rauscher, F.H. (2003). *Effects of piano, rhythm, and singing instruction on the spatial reasoning of at-risk children.* Proceedings of the European Society for the Cognitive Sciences of Music. Hannover, Germany: Hannover University Press.

Reitan, R.M., & Wolfson, D. (2004). Theoretical, methodological, and validational bases of the Halstead-Reitan Neuropsychological Test Battery. In G. Goldstein, S.R. Beers, & M. Hersen (Eds.), *Comprehensive handbook of psychological assessment* (pp. 105–131). Hoboken, NJ: Wiley.

Resnick, L.B. (1983). A developmental theory of number understanding. In H.P. Ginsburg (Ed.), *The development of mathematical thinking* (pp. 109–151). New York: Academic Press.

Resnick, L.B., Nesher, P., Leonard, F., Magone, M., Omanson, S., & Peled, I. (1989). Conceptual bases of arithmetic errors: The case of decimal fractions. *Journal for Research in Mathematics Education, 20,* 8–27.

Reuhkala, M. (2001). Mathematical skills in ninth-graders: Relationship with visuo-spatial abilities and working memory. *Educational Psychology, 21,* 387–399.

Reusser, K. (1988). Problem solving beyond the logic of things: Contextual effects on understanding and solving word problems. *Instructional Science, 17,* 309–338.

Reys, B., & Lappan, G. (2007). Consensus or confusion? The intended math curriculum in state-level standards. *Phi Delta Kappan, 88,* 676–680.

Rickard, T.C., Romero, S.G., Basso, G., Wharton, C., Flitman, S., & Grafman, J. (2000). The calculating brain: An fMRI study. *Neuropsychologia, 38,* 325–335.

Rittle-Johnson, B. (2006). Promoting transfer: Effects of self-explanation and direct instruction. *Child Development, 77,* 1–15.

Rittle-Johnson, B., & Kmicikewycz, A.O. (2008). When generating answers benefits arithmetic skill: The importance of prior knowledge. *Journal of Experimental Child Psychology, 101,* 75–81.

Rittle-Johnson, B., Siegler, R.S., & Alibali, M.W. (2001). Developing conceptual understanding and procedural skill in mathematics: An iterative process. *Journal of Educational Psychology, 93,* 346–362.

Rivera, S.M., Reiss, A.L., Eckert, M.A., & Menon, V. (2005). Developmental changes in mental arithmetic: Evidence for increased functional specialization in the left inferior parietal cortex. *Cerebral Cortex, 15,* 1779–1790.

Rivera-Batiz, F.L. (1992). Quantitative literacy and the likelihood of employment among young adults in the United States. *The Journal of Human Resources, 27,* 313–328.

Roberge, J.J., & Flexer, B.K. (1983). Cognitive style, operativity, and mathematics achievement. *Journal for Research in Mathematics Education, 14,* 344–353.

Robinson, C.S., Menchetti, B.M., & Torgesen, J.K. (2002). Toward a two-factor theory of one type of mathematics disability. *Learning Disabilities Research and Practice, 17,* 81–89.

Robinson, K.M., Arbuthnott, K.D., Rose, D., McCarron, M.C., Globa, C.A., & Phonexay, S.D. (2006). Stability and change in children's division strategies. *Journal of Experimental Child Psychology, 93,* 224–238.

Romberg, T.A., & Collins, C. (2000). The impact of standards-based reform on methods of research in schools. In A.E. Kelly & R.A. Lesh (Eds.), *Handbook of research design in mathematics and science education* (pp. 73–85). Mahwah, NJ: Lawrence Erlbaum Associates.

Rosenberger, P.B. (1989). Perceptual-motor and attentional correlates of developmental dyscalculia. *Annals of Neurology, 26,* 216–220.

Ross, P.A., & Baden, J.P. (1991). The effects of token reinforcement versus cognitive behavior modification on learning-disabled students' math skills. *Psychology in the Schools, 28,* 247–256.

Rossor, M.N., Warrington, E.K., & Cipolotti, L. (1995). The isolation of calculation skills. *Journal of Neurology, 242,* 78–81.

Rothman, R.W., & Cohen, J. (1989). The language of math needs to be taught. *Academic Therapy, 25,* 133–142.

Rotzer, S., Kucian, K., Martin, E., von Aster, M., Klaver, P., & Loenneker, T. (2008). Optimized voxel-based morphometry in children with developmental dyscalculia. *NeuroImage, 39,* 417–422.

Rourke, B.P. (1993). Arithmetic disabilities, specific and otherwise: A neuropsychological perspective. *Journal of Learning Disabilities, 26,* 214–226.

Rousselle, L., & Noël, M.-P. (2007). Basic numerical skills in children with mathematics learning disabilities: A comparison of symbolic vs. non-symbolic number magnitude processing. *Cognition, 102,* 361–395.

Roux, F.E., Boetto, S., Sacko, O., Chollet, F., & Tremoulet, M. (2003). Writing, calculating, and finger recognition in the region of the angular gyrus: A cortical stimulation study of Gerstmann syndrome. *Journal of Neurosurgery, 99,* 716–727.

Royer, J.M., & Garofoli, L.M. (2005). Cognitive contributions to sex differences in math performance. In A.M. Gallagher & J.C. Kaufman (Eds.), *Gender differences in mathematics* (pp. 99–120). New York: Cambridge University Press.

Rubinsten, O., & Henik, A. (2005). Automatic activation of internal magnitudes: A study of developmental dyscalculia. *Neuropsychology, 19,* 641–648.

Rubinsten, O., Henik, A., Berger, A., & Shahar-Shalev, S. (2002). The development of internal representations of magnitude and their association with Arabic numerals. *Journal of Experimental Child Psychology, 81,* 74–92.

Rudnitsky, A., Ethridge, S., Freeman, S. & Gilbert, T. (1995). Learning to solve addition and subtraction word problems through a structure-plus-writing approach. *Journal for Research in Mathematics Education, 26,* 467–486.

Rusconi, E., Walsh, V., & Butterworth, B. (2005). Dexterity with numbers: rTMS over the left angular gyrus disrupts finger gnosis and number processing. *Neuropsychologia, 43,* 1609–1624.

Santens, S., & Gevers, W. (2008). The SNARC effect does not imply a mental number line. *Cognition, 108,* 263–270.

Sarnecka, B.W., & Gelman, S.A. (2004). *Six* does not just mean *a lot:* Preschoolers see number words as specific. *Cognition, 92,* 329–352.

Saucier, D., Bowman, M., & Elias, L. (2003). Sex differences in the effect of articulatory or spatial dual task interference during navigation. *Brain and Cognition, 53,* 346–350.

Saucier, D.M., Green, S.M., Leason, J., MacFadden, A., Bell, S., & Elias, L.J. (2002). Are sex differences in navigation caused by sexually dimorphic strategies or by differences in the ability to use the strategies? *Behavioral Neuroscience, 116,* 403–410.

Schliemann, A.D., & Nunes, T. (1990). A situated schema of proportionality. *British Journal of Developmental Psychology, 8,* 259–268.

Schmidt, W. (2001, April 26). Testing in our schools [Television interview]. In *Frontline.* New York and Washington, DC: Public Broadcasting Service. Retrieved on September 20, 2008, from http://www.pbs.org/wgbh/pages/frontline/shows/schools/interviews/schmidt.html

Schmidt, W., Houang, R., & Cogan, L. (2002). A coherent curriculum: The case of mathematics. *American Educator, 26*(2), 1–18.

Schrank, F.A., Mather, N., McGrew, K.S., & Woodcock, R.W. (2003). *Woodcock-Johnson Diagnostic Supplement to the Tests of Cognitive Abilities.* Rolling Meadows, IL: Riverside Publishing.

Schunk, D.H., & Cox, P.D. (1986). Strategy training and attributional feedback with learning disabled students. *Journal of Educational Psychology, 78,* 201–209.

Schwartz, D.L., & Bransford, J.D. (1998). A time for telling. *Cognition and Instruction, 16,* 475–522.

Schwartz, D.L., & Moore, J.L. (1998). The role of mathematics in explaining the material world: Mental models for proportional reasoning. *Cognitive Science, 22,* 471–516.

Schwartz, S.H. (1971). Modes of representation and problem solving: Well evolved is half solved. *Journal of Experimental Psychology, 91,* 347–350.

Schwartz, S.H., & Fattaleh, D.L. (1972). Representation in deductive problem solving: The matrix. *Journal of Experimental Psychology, 95,* 343–348.

Scribner, S. (1986). Thinking in action: Some characteristics of practical thought. In R.J. Sternberg & R.K. Wagner (Eds.), *Practical intelligence: Nature and origins of competence in the everyday world* (pp. 13–30). New York: Cambridge University Press.

Seidenberg, M., Beck, N., Geisser, M., Giordani, B., Sackellares, J.C., Berent, S., et al. (1986). Academic achievement of children with epilepsy. *Epilepsia, 27,* 753–759.

Semple, J.S. (1992). *Semple math.* North Attleboro, MA: Stevenson Learning Skills.

Semrud-Clikeman, M., Biederman, J., Sprich-Buckminster, S., Lehman, B.K., Faraone, S.V., & Norman, D. (1992). Comorbidity between ADDH and learning disability: A review and report in a clinically referred sample. *Journal of the American Academy of Child and Adolescent Psychiatry, 31,* 439–448.

Seo, K.H., & Ginsburg, H.P. (2003). You've got to carefully read the math sentence . . .": Classroom context and children's interpretations of the equals sign. In A. Baroody & A. Dowker

(Eds.), *The development of arithmetic concepts and skills: Constructing adaptive expertise* (pp. 161–187). Mahwah, NJ: Lawrence Erlbaum Associates.

Shafrir, U., & Siegel, L.S. (1994). Subtypes of learning disabilities in adolescents and adults. *Journal of Learning Disabilities, 27,* 123–134.

Shalev, R.S., Auerbach, J., Manor, O., & Gross-Tsur, V. (2000). Developmental dyscalculia: Prevalence and prognosis. *European Child and Adolescent Psychiatry, 9,* II/58–II/64.

Shalev, R.S., Gross-Tsur, V., & Masur, D. (1995). Cognition, behavior and academic perform-ance in children with epilepsy. In S. Shinnar, N. Amir, & E. Branski (Eds.), *Childhood seizures* (pp. 170–178). Basel, Switzerland: Karger.

Shalev, R.S., Manor, O., Amir, N., & Gross-Tsur, V. (1993). Acquisition of arithmetic in normal children: Assessment by a cognitive model of dyscalculia. *Developmental Medicine and Child Neurology, 35,* 593–601.

Shalev, R.S., Manor, O., Amir, N., Wertman-Elad, R., & Gross-Tsur, V. (1995). Developmental dyscalculia and brain laterality. *Cortex, 31,* 357–365.

Shalev, R.S., Manor, O., & Gross-Tsur, V. (1997). Neuropsychological aspects of developmen-tal dyscalculia. *Mathematical Cognition, 3,* 105–120.

Shalev, R.S., Manor, O., & Gross-Tsur, V. (2005). Developmental dyscalculia: A prospective six year follow-up. *Developmental Medicine and Child Neurology, 47,* 121–125.

Shalev, R.S., Manor, O., Kerem, B., Ayali, M., Badichi, N., Friedlander, Y., et al. (2001). De-velopmental dyscalculia is a familial learning disability. *Journal of Learning Disabilities, 34,* 59–65.

Shapiro, B.J., & O'Brien, T.C. (1973). Quasi-child logics. *Educational Studies in Mathematics, 5,* 181–184.

Shaywitz, B.A., Shaywitz, S.E., Blachman, B.A., Pugh, K.R., Fulbright, R.K., Skudlarski, P., et al. (2004). Development of left occipitotemporal systems for skilled reading in children after a phonologically-based intervention. *Biological Psychiatry, 55,* 926–933.

Shaywitz, B.A., Shaywitz, S.E., Pugh, K.R., Mencl, W.E., Fulbright, R.K., Skudlarski, P., et al. (2002). Disruption of posterior brain systems for reading in children with developmental dyslexia. *Biological Psychiatry, 52,* 101–110.

Shaywitz, S.E., Escobar, M.D., Shaywitz, B.A., Fletcher, J.M., & Makuch, R. (1992). Evidence that dyslexia may represent the lower tail of a normal distribution of reading ability. *The New England Journal of Medicine, 326,* 145–150.

Shea, D.L., Lubinski, D., & Benbow, C.P. (2001). Importance of assessing spatial ability in in-tellectually talented young adolescents: A 20-year longitudinal study. *Journal of Educational Psychology, 93,* 604–614.

Shuman, M., & Kanwisher, N. (2004). Numerical magnitude and the human parietal lobe: Tests of representational generality and domain specificity. *Neuron, 44,* 557–569.

Shurtleff, H., Fay, G., Abbott, R., & Berninger, V.W. (1988). Cognitive and neuropsychologi-cal correlates of academic achievement: A levels of analysis assessment model. *Journal of Psychoeducational Assessment, 6,* 298–308.

Siegler, R.S. (1976). Three aspects of cognitive development. *Cognitive Psychology, 8,* 481–520.

Siegler, R.S. (1988). Strategy choice procedures and the development of multiplication skill. *Journal of Experimental Psychology: General, 117,* 258–275.

Siegler, R.S. (2003). Implications of cognitive science research for mathematics education. In J. Kilpatrick, W.G. Martin, & D.E. Schifter (Eds.), *A research companion to principles and stan-dards for school mathematics* (pp. 219–233). Reston, VA: National Council of Teachers of Mathematics.

Siegler, R.S., & Booth, J.L. (2004). Development of numerical estimation in young children. *Child Development, 75,* 428–444.

Siegler, R.S., & Booth, J.L. (2005). Development of numerical estimation. In J.I.D. Campbell (Ed.), *Handbook of mathematical cognition* (pp. 197–212). New York: Psychology Press.

Siegler, R.S., & Mu, Y. (2008). Chinese children excel on novel mathematics problems even before elementary school. *Psychological Science, 19,* 759–763.

Siegler, R.S., & Opfer, J.E. (2003). The development of numerical estimation: Evidence for multiple representations of numerical quantity. *Psychological Science, 14,* 237–243.

Siegler, R.S., & Ramani, G.B. (2007). Playing linear numerical board games promotes low-income children's numerical development. *Developmental Science, 11,* 655–661.

Siegler, R.S., & Svetina, M. (2006). What leads children to adopt new strategies? A micro-genetic/cross sectional study of class inclusion. *Child Development, 77,* 997–1015.

Sikora, D.M., Haley, P., Edwards, J., & Butler, R. (2002). Tower of London Test performance in children with poor arithmetic skills. *Developmental Neuropsychology, 21,* 243–254.

Silver, C.H., Pennett, D.-L., Black, J., Fair, G.W., & Balise, R.R. (1999). Stability of arithmetic disability subtypes. *Journal of Learning Disabilities, 32,* 108–119.

Simmons, F.R., & Singleton, C. (2006). The mental and written arithmetic abilities of adults with dyslexia. *Dyslexia, 12,* 96–114.

Simmons, F.R., & Singleton, C. (2008). Do weak phonological representations impact on arithmetic development? A review of research into arithmetic and dyslexia. *Dyslexia, 14,* 77–94.

Simon, O., Mangin, J.F., Cohen, L., LeBihan, D., & Dehaene, S. (2002). Topographical layout of hand, eye, calculation and language related areas in the human parietal lobe. *Neuron, 33,* 475–487.

Simon, T.J. (1997). Reconceptualizing the origins of number knowledge: A "non-numerical" account. *Cognitive Development, 12,* 349–372.

Simos, P.G., Breier, J.I., Fletcher, J.M., Foorman, B.R., Castillo, E.M., & Papanicolaou, A.C. (2002). Brain mechanisms for reading words and pseudowords: An integrated approach. *Cerebral Cortex, 12,* 297–305.

Sirigu, A., Cohen, L., Zalla, T., Pradat-Diehl, P., VanEeckhout, P., Grafman, J., et al. (1998). Distinct frontal regions for processing sentence syntax and story grammar. *Cortex, 34,* 771–778.

Skwarchuk, S.L., & Anglin, J.M. (2002). Children's acquisition of the English cardinal number words: A special case of vocabulary development. *Journal of Educational Psychology, 94,* 107–125.

Sluzenski, J., Newcombe, N.S., & Satlow, E. (2004). Knowing where things are in the second year of life: Implications for hippocampal development. *Journal of Cognitive Neuroscience, 16,* 1443–1451.

Smith, A.B., Taylor, E., Brammer, M., Toone, B., & Rubia, K. (2006). Task-specific hypoactivation in prefrontal and temperoparietal brain regions during motor inhibition and task switching in medication-naive children and adolescents with Attention Deficit Hyperactivity Disorder. *American Journal of Psychiatry, 163,* 1044–1051.

Smith, E.E., & Jonides, J. (1997). Working memory: A view from neuroimaging. *Cognitive Psychology, 33,* 5–42.

Sohn, M.H., Goode, A., Koedinger, K.R., Stenger, V.A., Fissell, K., Carter, C.S., et al. (2004). Behavioral equivalence, but not neural equivalence—Neural evidence of alternative strategies in mathematical thinking. *Nature Neuroscience, 7,* 1193–1194.

Soifer, L.H. (2005). Development of oral language and its relationship to literacy. In J.R. Birsh (Ed.), *Multisensory teaching of basic language skills* (2nd ed., pp. 43–81). Baltimore: Paul H. Brookes Publishing Co.

Sorby, S.A., & Baartmans, B.J. (2000). The development and assessment of a course for enhancing the 3-D spatial visualization skills of first year engineering students. *Journal of Engineering Education, 89,* 301–307.

Sowder, J. (1992). Estimation and number sense. In D.A. Grouws (Ed.), *Handbook of research on mathematics teaching and learning* (pp. 371–389). New York: Macmillan.

Sowell, E. (1989). Effects of manipulative materials in mathematics instruction. *Journal for Research in Mathematics Education, 20,* 498–505.

Spanos, G., Rhodes, N.C., Dale, T.C., & Crandall, J. (1988). Linguistic features of mathematical problem solving: Insights and applications. In R.R. Cocking & J.P. Mestre (Eds.), *Lin-*

guistic and cultural influences on learning mathematics (pp. 221–240). Mahwah, NJ: Lawrence Erlbaum Associates.

Spelke, E., Phillips, A., & Woodward, A.L. (1995). Infants' knowledge of object motion and human action. In D. Sperber, D. Premack, & A.J. Premack (Eds.), *Causal cognition: A multidisciplinary debate* (pp. 44–78). Oxford, United Kingdom: Clarendon Press.

Spelke, E.S., & Tsivkin, S. (2001). Language and number: A bilingual study. *Cognition, 78,* 45–88.

Stanescu-Cosson, R., Pinel, P., van de Moortele, P.-F., LeBihan, D., Cohen, L., & Dehaene, S. (2000). Cerebral bases of calculation processes: Impact of number size on the cerebral circuits for exact and approximate calculation. *Brain, 123,* 2240–2255.

Stanley, J.C., Benbow, C.P., Brody, L.E., Dauber, S., & Lupkowski, A. (1992). Gender differences on eighty-six nationally standardized aptitude and achievement tests. In N. Colangelo, S.G. Assouline, & D.L. Ambroson (Eds.), *Talent development, Vol. 1: Proceedings from the 1991 Henry B. and Jocelyn Wallace National Research Symposium on Talent Development* (pp. 42–65). Unionville, NY: Trillium Press.

Starkey, P., Spelke, E.S., & Gelman, R. (1990). Numerical abstraction by human infants. *Cognition, 36,* 97–127.

Steen, L.A. (2007). How mathematics counts. *Educational Leadership, 65*(3), 9–14.

Steeves, K.J. (1983). Memory as a factor in the computational efficiency of dyslexic children with high abstract reasoning ability. *Annals of Dyslexia, 33,* 141–152.

Stein, J. (2001). The neurobiology of reading difficulties. In M. Wolf (Ed.), *Dyslexia, fluency, and the brain* (pp. 3–21). Timonium, MD: York Press.

Steingold, E., Spelke, E., & Kittredge, A. (2003, July). *Linguistic cues influence acquisition of number words.* Paper presented at the 25th annual conference of the Cognitive Science Society, Cambridge, MA.

Stengel, C. (n.d.). *Quotes by Casey.* Retrieved August 26, 2008, from http://www.caseystengel .com/quotes_by.htm

Stern, C., & Stern, M.B. (1971). *Children discover arithmetic: An introduction to structural arithmetic* (2nd ed.). New York: Harper & Row.

Sternberg, R.J., & Weil, E.M. (1980). An aptitude × strategy interaction in linear syllogistic reasoning. *Journal of Educational Psychology, 72,* 226–234.

Stevens, D.A., Connell, M.W., Colvin, S., Schwartz, P., Pardi, R., & Pilgrim, B. (2003, August). *Designing educational software to improve cognitive abilities: Pilot study results.* Paper presented at the Interactive Technologies Conference on Training, Education, and Job Performance Improvement, Arlington, VA.

Stevens, D., & Schwartz, P. (2006). *Symphony math.* Retrieved July 5, 2008, from http://www .symphonylearning.com

Stevenson, H.W., & Stigler, J.W. (1992). *The learning gap: Why our schools are failing and what we can learn from Japanese and Chinese education.* New York: Touchstone.

Stevenson, H.W., Stigler, J.W., Lee, S.Y., Lucker, G.W., Kitamura, S., & Hsu, C.C. (1985). Cognitive performance and academic achievement of Japanese, Chinese, and American children. *Child Development, 56,* 718–734.

Stigler, J.W., & Hiebert, J. (1999). *The teaching gap: Best ideas from the world's teachers for improving education in the classroom.* New York: The Free Press.

Stigler, J.W., Lee, S.Y., & Stevenson, H.W. (1986). Digit memory in Chinese and English: Evidence for a temporally limited store. *Cognition, 23,* 1–20.

Stigler, J.W., & Perry, M. (1998). Mathematics learning in Japanese, Chinese, and American classrooms. In G.B. Saxe & M. Gearhart (Eds.), *Children's mathematics: New directions for child development* (pp. 27–54). San Francisco: Jossey-Bass.

Swanson, H.L., & Beebe-Frankenberger, M. (2004). The relationship between working memory and mathematical problem solving in children at risk and not at risk for math disabilities. *Journal of Educational Psychology, 96,* 471–491.

Swanson, H.L., & Jerman, O. (2006). Math disabilities: A selective meta-analysis of the literature. *Review of Educational Research, 76,* 249–274.

Sweller, J., & Cooper, G.A. (1985). The use of worked examples as a substitute for problem solving in learning algebra. *Cognition and Instruction, 2,* 59–89.

Tamm, L., Menon, V., & Reiss, A.L. (2006). Parietal attentional system aberrations during target detection in adolescents with Attention Deficit Hyperactivity Disorder: Event-related fMRI evidence. *American Journal of Psychiatry, 163,* 1033–1043.

Tang, Y., Zhang, W., Chen, K., Feng, S., Ji, Y., Shen, J., et al. (2006). Arithmetic processing in the brain shaped by cultures. *Proceedings of the National Academy of Sciences USA, 103,* 10775–10780.

Tartre, L.A. (1990). Spatial skills, gender, and mathematics. In E. Fennema & G. Leder (Eds.), *Mathematics and gender* (pp. 27–59). New York: Teachers College Press.

Taylor, H.G., Burant, C.J., Holding, P.A., Klein, N., & Hack, M. (2002). Sources of variability in sequelae of very low birth weight. *Child Neuropsychology, 8,* 163–178.

Teisl, J.T., Mazzocco, M., & Myers, G.F. (2001). Assessing the utility of kindergarten teacher ratings for predicting first grade academic achievement. *Journal of Learning Disabilities, 34,* 286–293.

Temple, C.M. (1989). Digit dyslexia: A category-specific disorder in developmental dyscalculia. *Cognitive Neuropsychology, 6,* 93–116.

Terepocki, M., Kruk, R.S., & Willows, D.M. (2002). The incidence and nature of letter orientation errors in reading disability. *Journal of Learning Disabilities, 35,* 214–233.

Terlecki, M.S., Newcombe, N.S., & Little, M. (2008). Durable and generalized effects of spatial experience on mental rotation: Gender differences in growth patterns. *Applied Cognitive Psychology, 22,* 996–1013.

Thompson, C.A., & Opfer, J.E. (2008). Costs and benefits of representational change: Effects of context on age and sex differences in symbolic magnitude estimation. *Journal of Experimental Child Psychology, 101,* 20–51.

Tohgi, H., Saitoh, K., Takahashi, S., Takahashi, H., Utsugisawa, K., Yonezawa, H., et al. (1995). Agraphia and acalculia after a left prefrontal (F1, F2) infarction. *Journal of Neurology, Neurosurgery, and Psychiatry, 58,* 629–632.

Tournaki, N. (2003). The differential effects of teaching addition through strategy instruction versus drill and practice to students with and without learning disabilities. *Journal of Learning Disabilities, 36,* 449–458.

Trbovich, P.L., & LeFevre, J.-A. (2003). Phonological and visual working memory in mental addition. *Memory & Cognition, 31,* 738–745.

Trites, R.L., & Fiedorowicz, C. (1976). Follow-up study of children with specific (or primary) reading disability. In R.M. Knights & D.J. Bakker (Eds.), *The neuropsychology of learning disorders* (pp. 41–50). Baltimore: University Park Press.

Turkeltaub, P.E., Gareau, L., Flowers, D.L., Zeffiro, T.A., & Eden, G.F. (2003). Development of neural mechanisms for reading. *Nature Neuroscience, 6,* 767–773.

Turner-Ellis, S.A., Miles, T.R., & Wheeler, T.J. (1996). Speed of multiplication in dyslexics and nondyslexics. *Dyslexia: An International Journal of Research and Practice, 2,* 121–139.

Ungerleider, L.G. (1995, November 3). Functional brain imaging studies of cortical mechanisms for memory. *Science, 270,* 769–775.

U.S. Department of Education, National Center for Education Evaluation and Regional Assistance. (2003). *Identifying and implementing educational practices supported by rigorous evidence: A user friendly guide.* Washington, DC: Author.

U.S. Department of Education, National Center for Education Statistics. (1997). *Profiles of students with disabilities as identified in NELS 88* (NCES 97-254). Washington, DC: Author.

U.S. Office of Education. (1977). Assistance to states for education for handicapped children: Procedures for evaluating specific learning disabilities. *Federal Register, 42,* 65082–65085.

Uttal, D.H., Fisher, J.A., & Taylor, H.A. (2006). Words and maps: Developmental changes in mental models of spatial information acquired from descriptions and depictions. *Developmental Science, 9,* 221–235.

Uttal, D.H., Scudder, K.V., & DeLoache, J.S. (1997). Manipulatives as symbols: A new perspective on the use of concrete objects to teach mathematics. *Journal of Applied Developmental Psychology, 18,* 37–54.

Vandenberg, S.G., & Kuse, A.R. (1978). Mental rotation, a group test of three-dimensional spatial visualization. *Perceptual and Motor Skills, 47,* 599–604.

Vander Linde, L.F. (1964). Does the study of quantitative vocabulary improve problem solving? *Elementary School Journal, 65,* 143–152.

Vander Will, C. (1976). *The wording of spoken instructions to children and its effect on their performance of tasks* [Abstract]. Retrieved on April 25, 2007, from http://eric.ed.gov/ERICWebPortal/recordDetail?accno=EJ148969.

Van Dooren, W., DeBock, D., Hessels, A., Janssens, D., & Verschaffel, L. (2004). Remedying secondary school students' illusion of linearity: A teaching experiment aiming at conceptual change. *Learning and Instruction, 14,* 485–501.

van Galen, M.S., & Reitsma, P. (2008). Developing access to number magnitude: A study of the SNARC effect in 7- to 9-year-olds. *Journal of Experimental Child Psychology, 101,* 99–113.

Van Haneghan, J., Barron, L., Young, M., Williams, S., Vye, N., & Bransford, J. (1992). The *Jasper* series: An experiment with new ways to enhance mathematical thinking. In D.F. Halpern (Ed.), *Enhancing thinking skills in the sciences and mathematics* (pp. 15–38). Mahwah, NJ: Lawrence Erlbaum Associates.

Varley, R.A., Klessinger, N.J.C., Romanowski, C.A.J., & Siegal, M. (2005). Agrammatic but numerate. *Proceedings of the National Academy of Sciences USA, 102,* 3519–3524.

Varley, R., & Siegal, M. (2000). Evidence for cognition without grammar from causal reasoning and 'theory of mind' in an agrammatic aphasic patient. *Current Biology, 10,* 723–726.

Venkatraman, V., Soon, C.S., Chee, M.W., & Ansari, D. (2006). Effects of language switching on arithmetic: A bilingual fMRI study. *Journal of Cognitive Neuroscience, 18,* 64–74.

Vinckier, F., Naccache, L., Papeix, C., Forget, J., Hahn-Barma, V., Dehaene, S., et al. (2006). "What" and "where" in word reading: Ventral coding of written words revealed by parietal atrophy. *Journal of Cognitive Neuroscience, 18,* 1998–2012.

Vingerhoets, G., Santens, P., Van Laere, K., Lahort, P., Dierckx, R.A., & DeReuck, J. (2001). Regional brain activity during different paradigms of mental rotation in healthy volunteers: A positron emission tomography study. *NeuroImage, 13,* 381–391.

Volkow, N.D., Wang, G.-J., Fowler, J.S., Telang, F., Maynard, L., Logan, J., et al. (2004). Evidence that methylphenidate enhances the saliency of a mathematical task by increasing dopamine in the human brain. *American Journal of Psychiatry, 161,* 1173–1180.

von Aster, M. (2000). Developmental cognitive neuropsychology of number processing and calculation: Varieties of developmental dyscalculia. *European Child and Adolescent Psychiatry, 9,* II/41–II/57.

Voyer, D., Voyer, S., & Bryden, M.P. (1995). Magnitude of sex differences in spatial ability: A meta-analysis and consideration of critical variables. *Psychological Bulletin, 117,* 250–270.

Wakefield, D.V. (2000). Math as a second language. *The Educational Forum, 64,* 272–279.

Walsh, V., & Butler, S.R. (1996). The effects of visual cortex lesions in the perception of rotated shapes. *Behavioural Brain Research, 76,* 127–142.

Warren, E. (2003). Language, arithmetic and young children's interpretations. *Focus on Learning Problems in Mathematics, 25,* 22–35.

Warrington, E.K. (1982). The fractionation of arithmetic skills: A single case study. *Quarterly Journal of Experimental Psychology, 34,* 31–51.

Warrington, E.K., & Shallice, T. (1980). Word-form dyslexia. *Brain, 103,* 99–112.

Webb, R.M., Lubinski, D., & Benbow, C.P. (2007). Spatial ability: A neglected dimension in talent searches for intellectually precocious youth. *Journal of Educational Psychology, 99,* 397–420.

Wechsler, D. (1997). *Wechsler Adult Intelligence Scale—Third Edition (WAIS-III): Administration and scoring manual.* San Antonio, TX: Harcourt Assessment.

Wechsler, D. (2003). *Wechsler Intelligence Scale for Children—Fourth Edition (WISC-IV): Administration and scoring manual.* San Antonio, TX: Harcourt Assessment.

Welsh, M.C., Pennington, B.F., & Groisser, D.B. (1991). A normative-developmental study of executive function: A window on prefrontal function in children. *Developmental Neuropsychology, 7,* 131–149.

Wendt, P.E., & Risberg, J. (1994). Cortical activation during visual spatial processing: Relation between hemispheric asymmetry of blood flow and performance. *Brain and Cognition, 24,* 87–103.

Wheeler, L.J., & McNutt, G. (1983). The effect of syntax on low-achieving students' abilities to solve mathematical word problems. *Journal of Special Education, 17,* 309–315.

Whetstone, T. (1998). The representation of arithmetic facts in memory: Results from retraining a brain-damaged patient. *Brain and Cognition, 36,* 290–309.

Willis, J. (2007). *Brain-friendly strategies for the inclusion classroom: Insights from a neurologist and classroom teacher.* Alexandria, VA: Association for Supervision and Curriculum Development.

Wilson, A., Dehaene, S., Pinel, P., Revkin, S., Cohen, L., & Cohen, D. (2006). *Principles underlying the design of The Number Race, an adaptive computer game for remediation of dyscalculia.* Retrieved February 5, 2007, from http://www.behavioralandbrainfunctions.com/content/2/1/19

Wilson, A., Revkin, S., Cohen, D., Cohen, L., Dehaene, S. (2006). *An open trial assessment of The Number Race, an adaptive computer game for remediation of dyscalculia.* Retrieved February 5, 2007, from http://www.behavioralandbrainfunctions.com/content/2/1/20

Wilson, K., & Swanson, H.L. (2001). Are mathematics disabilities due to a domain-general or domain-specific working memory deficit? *Journal of Learning Disabilities, 34,* 237–248.

Wilson, W.S., & Naiman, D.Q. (2004). K-12 calculator usage and college grades. *Educational Studies in Mathematics, 56,* 119–122.

Witelson, S.F., Kigar, D.L., & Harvey, T. (1999). The exceptional brain of Albert Einstein. *The Lancet, 353,* 2149–2153.

Wolf, M., Bowers, P.G., & Biddle, K. (2000). Naming-speed processes, timing, and reading: A conceptual review. *Journal of Learning Disabilities, 33,* 387–407.

Wolf, M., & Denckla, M.B. (2005). *Rapid Automatized Naming and Rapid Alternating Stimulus (RAN/RAS) Tests.* Austin, TX: PRO-ED.

Wolfgang, C.H., Stannard, L.L., & Jones, I. (2001). Block performance among preschoolers as a predictor of later school achievement in mathematics. *Journal of Research in Childhood Education, 15,* 173–181.

Woodcock, R.W., McGrew, K.S., & Mather, N. (2001). *Woodcock-Johnson III Tests of Cognitive Abilities (WJ III).* Rolling Meadows, IL: Riverside Publishing.

Woodward, J. (2006). Developing automaticity in multiplication facts: Integrating strategy instruction with timed practice drills. *Learning Disability Quarterly, 29,* 269–289.

Wynn, K. (1998). Psychological foundations of number: Numerical competence in human infants. *Trends in Cognitive Science, 2,* 296–303.

Xin, Y.P., Jitendra, A.K., & Deatline-Buchman, A. (2005). Effects of mathematical word problem-solving instruction on middle school students with learning problems. *The Journal of Special Education, 39,* 181–192.

Yin, W.G., & Weekes, B.S. (2003). Dyslexia in Chinese: Clues from cognitive neuropsychology. *Annals of Dyslexia, 53,* 255–279.

Zago, L., Pesenti, M., Mellet, E., Crivello, F., Mazoyer, B., & Tzourio-Mazoyer, N. (2001). Neural correlates of simple and complex calculation. *NeuroImage, 13,* 314–327.

Zawaiza, T.R.W., & Gerber, M.M. (1993). Effects of explicit instruction on math word-problem solving by community college students with learning disabilities. *Learning Disability Quarterly, 16,* 64–79.

Zebian, S. (2005). Linkages between number concepts, spatial thinking, and directionality of writing: The SNARC effect and the REVERSE SNARC effect in English and Arabic monoliterates, biliterates, and illiterate Arabic speakers. *Journal of Cognition and Culture, 5,* 165–190.

Zelazo, P.D., Carter, A., Reznick, J.S., & Frye, D. (1997). Early development of executive function: A problem-solving framework. *Review of General Psychology, 1,* 198–226.

Zentall, S.S. (1993). Research on the educational implications of attention deficit hyperactivity disorder. *Exceptional Children, 60,* 143–153.

Zentall, S.S. (2007). Math performance of students with ADHD: Cognitive and behavioral contributors and interventions. In D.B. Berch & M.M.M. Mazzocco (Eds.), *Why is math so hard for some children? The nature and origins of mathematical learning difficulties and disabilities* (pp. 219–243). Baltimore: Paul H. Brookes Publishing Co.

Zettle, R.D. (2003). Acceptance and commitment therapy (ACT) vs. systematic desensitization in treatment of mathematics anxiety. *The Psychological Record, 53,* 197–215.

Zorzi, M., Priftis, K., & Umiltà, C. (2002). Brain damage: Neglect disrupts the mental number line. *Nature, 417,* 138–139.

Zur, O., & Gelman, R. (2004). Young children can add and subtract by predicting and checking. *Early Childhood Research Quarterly, 19,* 121–137.

Index

Page numbers followed by *f* indicate figures.